The
Call Center
Handbook

5th Edition

BY KEITH DAWSON

CMP**Books**

THE CALL CENTER HANDBOOK
Published by CMP Books
An Imprint of CMP Media LLC
Main Office: CMP Books, 600 Harrison St., San Francisco, CA 94107 USA
Phone: 415-947-6615; FAX: 415-947-6015

CMP
United Business Media

ISBN 1-57820-305-8
For individual orders, and for information on special discounts for quantity orders, please contact:
CMP Books Distribution Center
6600 Silacci Way, Gilroy, CA 95020
Tel: 800-500-6875 or 408-848-5784; Fax: 408-848-5784
Email: cmp@rushorder.com; web: www.cmpbooks.com

Cover and Text Design by Robbie Alterio

Distributed to the book trade in the U.S. by
Publishers Group West, 1700 Fourth St., Berkeley, CA 94710

Distributed in Canada by:
Jaguar Book Group, 100 Armstrong Avenue, Georgetown, Ontario M6K 3E7 Canada

Manufactured in the United States of America.

04 05 06 07   5 4 3 2 1

Table of Contents

Part VI - Management & Operations 211

Part VII - Outside the Center 247

Why Call Centers Still Matter

Chances are you have this book in your hand because you run, or work in, or have something to do with, a call center. Chances are also that when people outside work ask you what you do, you have to explain to them exactly what a call center is, and they don't always get it. They focus on the things that they dislike about call centers — those telemarketing calls they get at dinnertime or when they're putting the baby to bed. Or the long time they spend on hold trying to get someone to explain why there's an error on their bank statement.

Sometimes call centers get pretty bad press — especially when the economy turns sour and jobs are shipped offshore. And for every story we can tell you about the people who work long and hard to make their centers the epitome of good service, what many people remember is the rude call, or the rare time when the rep simply can't fix whatever has gone wrong.

Obviously there's more to call centers than the face they present to the consumer. But that's the reality — that people interact with call centers all the time, and often come away frustrated and bewildered. That's one of the main reasons this book exists: to help you improve the way the call center operates, and following that, to improve the relationship between your company and your customers.

This book is for anyone who works in a call center. For anyone who sells by phone. Or who helps customers. It's about all the stuff that's used in call centers, the technology, hardware and software. It's also about the kinds of services that call centers buy, things like toll free and long distance services, outsourced call center help, site selection assistance and consulting. And it's about the people who work in centers: how to keep reps happy, interested, well trained and excited about their jobs. How to make sure that you don't spend a fortune in training only to lose those people after a few short months because some preventable thing you're doing is driving them away.

A call center is a traditionally defined as a physical location where calls are placed, or received, in high volume for the purpose of sales, marketing, customer service, telemarketing, technical support or other specialized business activity. One early definition described a call center as a place of doing business by phone that combined a centralized database with an automatic call distribution system. That was pretty good for 1985, but today it's more than that. These days we expand the definition in two directions.

First, we expand it to include call-taking and call-making organizations that were originally overlooked, like fundraising and collections organizations, and help desks, both internal and external.

And more controversially, we expand it to include centers that handle more than the traditional voice call — lets call them call centers plus. These would be centers that handle voice plus fax, or email, live Web chat centers, video interactions — all the many real and hypothetical customer interactions that are now possible.

Estimates of the number of call centers in North America range from 20,000 to as high as 200,000. The reality is probably somewhere around 120,000 depending on what you consider a call center. Some experts believe that you shouldn't count centers below a certain number of agents (or "seats"). I believe in the widest possible definition, all the way down to micro-centers of four or five people. Why? Because those centers face many of the same kinds of problems on a daily basis as their larger cousins: problems of training, staffing, call handling, technology assessment, and so on. Those smaller centers have to put the same kind of face forward to the customer as the largest centers, in order to remain competitive. And more often than not, those small center become medium-sized centers over time.

Call centers are generally set up as large rooms, with workstations that include a computer, a telephone set (or headset) hooked into a large switch and one or more supervisor stations. It may stand by itself, or be linked with other centers. It will most likely be linked to a corporate data network, including mainframes, microcomputers and LANs.

Call centers were first recognized as organized corporate departments around 30 years ago, in their largest incarnations: airline reservation centers, catalog ordering companies, problem solvers like the GE Answer Center. Until the early 1990s, only the largest centers could afford the investment in technology that allowed them to handle huge volumes - this was the development of the automatic call distributor, which in those days was a customized, proprietary phone system designed to funnel as many calls as possible into a single large site. (In fact, apocryphal industry legend likes to point to the installation of Rockwell's first Galaxy ACD at Continental Airlines 30 years ago as the de facto beginning of "modern" times.)

More recently, with the development of PC LANs, client/server software systems, open phone systems and browser-based desktops for data access, any call center can have an advanced call handling and customer management system, even down to ten agents or less.

As companies have learned that service is the key to attracting and maintaining customers (and hence, revenue), the common perception of the call center has changed. It is rarely seen as a luxury anymore. In fact, it is often regarded as a competitive weapon. In some industries (catalog retailing, financial services, hospitality) a call center is the difference between being in business and not being in business. In other industries (cable television, utilities) call centers have been the centerpiece of corporate attempts to quickly overhaul service and improve their image.

It's a good working hypothesis to assume that any company that sells any product has a call center, or will shortly have one, because it is the most effective way to reach (and be reached by) customers.

Just when you thought you knew what you were doing — technology is redefining the call center, changing it into something bigger, more complex, and ultimately more customer-pleasing.

You could choose to define a center in terms of its physical reality, like the traditional definitions just given. It is a roomful of people, devoted to the task of making and/or receiving calls to and from customers. It is the place where those calls are handled, and the accumulation of technologies that assist: phone lines, switches, software, human expertise.

Or, you could look at it from the point of view of function — the call center as the locus for customer satisfaction. In that view, the center is the "place" where the customer goes to complain, to place an order, or get help — even if the agents are widely disconnected from each other, or if the database is halfway around the world from them.

There may not seem to be too much difference between those two points of view. Until recently, there wasn't. Whether you subscribed to one or another definition mattered little in the day to day running of a center. You could even hold both views without much cognitive dissonance.

Call centers have emerged as powerful, strategic tools in the fight to gain and keep customers. Running a center has become its own corporate discipline. The call center industry has become an industry — not simply a collection of dissociated vertical markets with similar needs. We've seen phenomenal growth in all segments — equipment sales, outsourcing services, toll-free traffic, customer sales made by phone.

And yet, the moorings are coming loose. Something ironic has happened. At almost the precise moment that we can herald the arrival of the call center industry, we can see, looming out there on the horizon, the signs of something coming along to replace what we know as the "call center."

The physical center itself is devolving. Smaller centers are more practical. You can put a fifty seat center into virtually any town or city in the US, without worrying about telecom infrastructure or labor. Cities, states and most recently, foreign countries are falling over each other to offer tax incentives to attract call centers.

Centers can use agents-at-home, virtual agents who sign into a center from their homes whenever demand requires, the ultimate in just-in-time staffing. I heard of one company that trains spouses of call center agents. Those spouses, who already know much about the company, are then equipped to pick up part-time work on very short notice, and can sign into the call center from their kitchen tables.

This is happening more and more. Technology makes the role of an agent more powerful — agenting is more of an analytic and interpretive skill, as well as more interpersonal. They have access to more information about the customer and the company. And the kinds of questions they are called upon to answer are different. They are higher level, more complex, often requiring more decision-making authority for customer service and support.

With so many alternate entry points into a company's sales/service operation, we need to rethink the traditional measures of service level, revenue generated per call, cost per call, and so on. What is the relative cost of a Web hit, and its benefit? In the same vein, can we afford to treat email requests for service any differently than we do live calls?

Most of what we now describe as a call is really best described as a transaction or an interaction between two parties, you and the customer. Eventually most customer interactions won't involve an agent. They will be electronically processed database transactions.

We have to recognize that the call center is more than the simple place we defined earlier, a place for making and taking calls. It is best described by its function — as a collection of people and technologies whose role is to serve customers.

You are on one side of a chasm, your customers on the other side. There are many ways to get from one side to the other. The customer chooses which route depending on what he needs.

There is already evidence that the role of call center agent is changing, into more of a knowledge worker. Reps are now, more than ever, on a career track to supervisory and management positions.

Things are changing faster than ever before. As soon as this book hits your hands, we'll be ready with a pile of new technologies to write about, new products that are out there, and new ideas about how to run the call center.

As always with any project this size, you might find errors of fact or judgment in this book. If so, please let me know, and if possible I'll correct them in the next edition.

— Keith Dawson
Winter, 2004

Part One

The Physical Center

You can place a call center anywhere you like. I've seen centers in office buildings, strip malls and industrial parks; squeezed into back rooms and former warehouses; converted supermarkets and trading floors — there's no end to the ways to house a call center.

Chapter One

How Call Centers Evolve, or, How to Start Putting Your Center Into Perspective

When you want to make generalizations and spot trends, one of the most effective things to do is to try to make a model that simulates the broadest possible range of circumstances and see if that model acts like real life.

Over the years, I've had just such a working model for how call centers evolve over time. I've had to refine it, but I think that even today it describes with a fair degree of accuracy the struggles that call center has to go through to deliver better service at a consistently high level of efficiency and lower cost.

The reason you create a model is to try exploring the general principles that guide a process, in this case, the process of service delivery. Since I created the Six-Stage Model, much has changed in the outside world that affects how call centers operate, and how they change over time. As you might expect, it's the later stages of call center development that are the most interesting, and that incorporate leading edge techniques and technologies. It's where you find people grappling with the most complex and nuanced issues that face call centers.

My Six-Stage Model of Call Center Development has less to do with what technologies are used and more with the way the center interacts with the rest of the company. Here's how it looks (and keep in mind that these are broad-brush descriptions rather than definitive categories):

Stage 1: **Startup**. Also known as the informal or departmental call center mode. Organizationally, you find this in either small, growing companies that have not yet created a structure for service delivery, or in larger ones that allow diffuse, fractured approaches to spring up ad hoc.

This is one of the most haphazard steps. It's the point at which a company is operating with little or no strategic view of the value of customers, or the consequences of good or bad service delivery. In this environment you might find a marketing department answering calls, or a voice mail box set up to handle customer inquiries. What you won't find is any kind of measurement, or coordination, or support from the higher levels of the company. You also won't find any specific service-related technology investment. Instead, you find people on the fringe of the organization trying to make do with existing tools like PBXs.

The people doing the call handling are not, at this stage, professional reps. They are low level, perhaps entry level people filling in where gaps are found. Unfortunately, in this case service delivery can be seen as one large gap.

Stage 2: **Triage**. At some point someone notices that something has to be done. You have to respond in some way to the customer base — even the upper managers know this. But often their first response is an inadequate one. It is reactive, and often guided by a single horrible metric like hold time, or products being returned in droves.

The response usually takes the form of traditional call center tools: ACD features added to a PBX, even a standalone ACD, and an organizing of the service-delivery hierarchy. Someone is tasked with addressing the problem; that someone is given a budget and a mandate and little else. This is the first step toward creating the call center as we know it today, an infrastructure built to receive and handle customer inquiries.

Often, the problems that drove the process in the first place can be alleviated (note that we didn't say solved) through the application of tools that are very mature and easily controlled.

Stage 3: **The Organized Center**. The people in charge become more professional, gaining a better grasp of the array of tools at their disposal, and the pros and cons of each one. The agents themselves also become professionalized; it's at this stage that training, monitoring, and incentives become factors in agent management.

The call center management is still focused on cost control, usually at the behest of corporate policy planners. Here's where you start to see some of the more interesting technology make its appearance, as managers try to squeeze operations to make them more productive in response to the fundamental call center paradox: the need to simultaneously cut costs and improve service. You see attention paid to all the cheap, reliable tech that's become very standard, like front-end IVR processing and the first glimmer of the possibilities of bringing the telecom infrastructure together with the database. (We used to call this CTI or computer telephony — now it makes more sense to lump it all together into CRM.)

And of course you see a lot more attention paid to reporting and analysis at this stage. But that analysis is mainly at the level of the call and the agent — not the customer.

It's also at this stage that some of the newer, more interesting technologies start making their appearance. Workforce management software, for example, is often brought in here to make the scheduling of agents more cost-effective and fairer.

Stage 4 is one of **Continuous Improvement**. That's the point at which the tension between the external needs of the company and the internal needs of the center are at their most apparent. From outside comes the imperative to cut costs, as before, but also comes the need for more real, useful information about what's happening with customers. This is where you begin to see the CRM mentality creeping in — not always emerging from within the center. Just as often it comes from the need outside departments have for organized, coordinated information relating call center activity to what the rest of the company is doing in marketing, product development, shipping, financial analysis, and so forth.

It's also the stage at which call centers become really rigorous internally about things like monitoring and training programs, about workforce management software, and sometimes begin experimenting with things like skills-based routing and simulation.

In Stage 5, the forces at work in Stage 4 have matured a bit, and as a result, outsiders at the organization have come to see the call center more as a **Strategic Asset** than as a money pit, or a drag on the rest of the company.

At this point, people are starting to cast about for a set of metrics they can use to define the relationship between the company and the customer; often the best they can do is to frame that relationship in terms of call center stats — number of calls, how well satisfied those customers are, and the nexus between customer value and customer cost.

And then we come to Stage 6, a magical point at which the importance of the physical call center is reduced and is replaced by the company-customer relationship as the ongoing focal point. I call this the **Mass Customization of Service**.

At this point, the call center and its company become a tightly integrated whole. Think of it this way: in a call center this highly evolved, each rep should theoretically have at his or her disposal all the information needed to handle the customer at the point of interaction — no more or less information than was needed, at the precise moment of contact. And the caller and rep should be as precisely matched as possible, with the agent empowered to do whatever necessary to meet the customer's needs, within the context of knowing what business rules to apply, and what the value of the customer is.

Stage 6 is a) not here yet in any meaningful way, b) a theoretical end point at which the boundaries between call center and company are indistinguishable, and c) a largely unattainable goal based on the technology and business contingencies of the day.

That's all well and good, and one can make a very nice living moving a call center down the road from Stage 1 or 2 through to 5 or 6. But I've found that it doesn't always apply — there are a couple of sideways paths that call centers can take, alternative developmental options that are increasingly common.

First, the dot-com online retailing boom and bust cycle we've just gone through revealed some things about online customer service. Mainly, that it isn't any good. Small companies that came out of the Internet industry sometimes thought that Internet-based tools for service delivery were adequate for their customers — that customers would be satisfied asking for help by perusing a Web-page and sending an email. Many invested in front-end email processing tools, and even a backend customer management system, only to neglect the telephone infrastructure. And people being people, they do still like to pick up the phone and call when something goes wrong (especially when it goes wrong with Christmas presents ordered online.)

So that first side stage involves what I call **Backpedaling to the Phones**. It's a stage that could occur in any industry, really, that set itself up to handle customer interactions through non-telephony modes, and belatedly realizes that the customers couldn't care a whit about a high-tech Internet service strategy but demand satisfaction by phone.

In these cases, though, the company doesn't backpedal all the way to Startup or Triage anymore. They go full bore into Stage 3. That's because unlike ten years ago, or even five, it's reasonably fast to throw together a technically robust call-handling infrastructure using what are now mature, proven, well-integrated tools.

The second side stage is more of a decision point that a lot of call centers find themselves reaching very early in their life cycle nowadays: **Multichannel Choices**: What do you do about emails? Or Web interaction? Before you even get to the question of HOW, all call centers face the question of whether to integrate these alternate modes of customer interaction into the telephony center at all. Which road you go down influences which set of technologies you're going to buy, and what face you're going to present to the customer, and how your internal management will be structured, and how much you'll spend to integrate the different channels, and even how you train agents. It's immensely important, and it's something that's sidestepped all the time, because it's so profoundly complicated, and affects so many different aspects of business planning, that it's often just decided by accident or default or both, often from outside the center. We're not talking about the technology decision here — we're talking just about the raw gut choice of do we handle emails in the call center, or do we create a separate center to handle them? Do we make telephone reps lead Internet chats, or leave it to a separate group?

Because it is so hellishly complicated, a lot of companies head straight for the relative safety of the third side stage: **Outsource the Whole Mess**. This is a fairly clear proposition, especially when there's no industrywide operational consensus for how best to run a center under multi-channel conditions. And when it's clear that what you're looking at buying is a set of very expensive and very transitional technologies that are going to look quite different five years hence. More than at any time since we've been covering call centers, outsourcing all or part of these multichannel components and the CRM overhead is a popular way to cut uncertainty's Gordian knot.

Obviously this model isn't going to look right to everyone; not all centers are created equal, and not all company agendas are the same. I'm just trying to paint (with a very broad brush) some of the common choices made by people grappling with similar problems. As things get more complicated, though, and as the modes used to connect with customers multiply, the "call center" will be harder and harder to describe in a universal way.

As you read (or leaf through) this book, try to keep these pathways in mind, and try to picture where your center fits on this timeline. It will help you decide which mix of technologies and operations is right for you.

Destination: Call Center

Before you buy a switch, hire staff, install a single piece of software, train any agents — before you take a single call - you're going to have to find a place to put your center. Site selection is one of the biggest decisions you'll make. It affects every decision that comes after — from the kinds of wiring you install to the services you offer your customers, for years to come. Where do you locate your call center to provide the maximum advantage to your organization and your customers?

Finding a site for your call center is more difficult than finding a site for many other types of business. And it is more difficult to find the ideal site for your call center now than it was just a few years ago.

The choice of locations has never been broader — you can put an effective center just about anywhere these days. But having more choices means you have to do your homework better than ever to find the right site for your center. It used to be a simple matter. You'd put your center in an established call center-friendly city, like Omaha, hoping to get the best telecom connections, and an educated, accent-free workforce.

But that's not the most cost-effective solution anymore. While there are a huge number of centers scattered throughout the American (and Canadian) Midwest, other factors let you enlarge the area of consideration. Many companies no longer see the Midwest as an attractive area for new centers (though existing ones continue to thrive) because of the tight labor market.

In essence, the traditional barriers have been eliminated, which makes the decision of where to locate if anything, harder. You can get good telecommunications services just about anywhere in the continental US and Canada. Similar conditions exist in other regions of the developed world. And worldwide the cost of telecom has dropped dramatically through the past decade.

A call center is not physically demanding as to the kind of space you put it in, either. I've centers in office buildings, strip malls and industrial parks; squeezed into back rooms and former warehouses; converted supermarkets and trading floors — there's no end to the ways to house a call center.

Given that, you can place a call center anywhere you_ like. It means that you can choose the physical surroundings based on your specific company's needs. If you need to be near an order processing center, for example, or the CEO's summer cottage, you'll not find that too much of a problem.

What is good news for call center managers is there are plenty of locations, both in the United States and abroad, that are eager to have your call center join their business community.

▇ Selection Criteria

Labor. "Call centers have traditionally drawn from the marginally employed," says call center expert Madeline Bodin. "People like housewives, students with flexible schedules, people who have to work ·part time, or recent graduates. With unemployment so low, people have other choices." She says that given the choice between a job in a call center and something else, there has to be a reason for potential employees at that level going to choose the center. Often, there isn't one.

Kevin Leonard of Strategic Outsourcing Corp. specializes in hiring in bulk — hundreds of agents at a time for large, high-end call centers in the southwest. "What most call center people do is call temp agencies," he says. "The signal is sent to all the temp agencies and they find everyone they can't place anywhere else." When you staff by the hundred in a single concentrated area, especially under time pressure, the call center management ends up under the gun, with no real opportunity to check the skills of the pool as a whole.

What this means for call center site selection is that the old rules don't necessarily apply. Going to a location because of the cost of labor doesn't make sense. Economic conditions will change over time, changing the cost of labor. What you need to look for is the quality and continuity of a region's labor force. Quality, meaning people are well educated, with computer and communication skills, and a solid work ethic. Continuity, of course, meaning that the labor pool is deep enough to supply a constant stream of workers, even at your highest projected rate of turnover.

While many believe that an area's unemployment rate is the most important indicator of workers available, the best indicator is how likely that market is to grow. Unemployment gives you a quantity of workers from which to choose, but is not a guarantee of quality.

If the labor market is growing, a company can set high standards for recruiting. Growth allows that company to let go employees who don't meet its standards. High growth also lessens the inflationary pressures on wages and benefits in that market. In the US, this kind of growth is usually indicated by a high number of people moving to the area. A large percentage of young people — and even a high birth rate — are other signs to watch for. Salt Lake City, for example, is a recently a popular call center location, in part because of its nation-topping birth rate.

A growing labor pool is more important for call centers than for most other industries. Because of the stress and the high turnover rate, call centers tend to burn out a labor pool faster. After a while your help wanted ads may go unanswered by a populace battle-scarred by call center work.

A steady supply of new workers can make finding employees easier when your best employee retention efforts fail. Also, call center employees are different from most. They are usually part-time workers that don't share a demographic profile with the usual nine-to-fivers.

Analyze an area for students, senior citizens, housewives, people in a career transition and people new to the area. It is from these labor pockets that call centers usually draw their workers.

Educational institutions. As noted above, students are historically a ready source of call center reps. Placing a center in an area that has one or more colleges or even a single large university can give you an ongoing flow of admittedly transient workers. They won't stay around long, but they are often available on short notice, and they can work odd shifts.

Increasingly, vocational education is a powerful draw for call centers. Community colleges are beginning to offer certification programs in the kinds of skills that are necessary for call center work.

When rating an area, look at what the educational infrastructure is like: does the community support it? Is it channeling people into high tech jobs, and preparing them for entry level high tech work (which is what call centers are, really) with the appropriate mix of people skills and technical skills? Are graduates staying in the community, or are they leaving for something else? And is there another dominant industry in the community that's soaking up the newly graduated workforce? An industry that's cyclical might create boom and bust cycles of high unemployment that will affect your cost structure — not necessarily always badly, but it's something you have to think about going into an area.

Using student labor is a very personal corporate decision. Students offer a ready source of low-cost labor, have higher academic credentials and are generally articulate. But to students, school comes first, employment is second. Their schedules are more important than their jobs.

Companies that can compensate for this, and students' high turnover rate, find a ready source of student labor very attractive. Other companies avoid student workers.

Whether you are looking for students or graduates, junior colleges, trade schools, business schools and even high schools are often-overlooked resources. Some schools even fashion their curriculum to meet the needs of business, with "call center certification" a growing trend in some areas.

Real estate. Outside the major cities, this is one of the things, oddly, you have to worry about least. Space for call centers is plentiful, and customizable to your heart's content.

What you should look for is outside the scope of this book because it's going to depend entirely on the kind of call center you are creating, and the way that center relates to the rest of your company. For example, company culture will determine the likelihood that the center will add seats over time, perhaps doubling or tripling in size. Or alternatively, if the company has a bias to opening multiple centers and keeping them small. There are advantages to either method, but again, it depends on internal factors more than external.

People used to stay away from cities like New York and Los Angeles because the cost of office space was so prohibitively high (as are the taxes). In recent years, though, cities (and their near suburbs) are more popular for some industries despite the high real estate costs. Some financial services businesses, for example, want their call centers closer to their trading desks and corporate headquarters, particularly when the call center is used to respond to Internet inquiries as well as voice calls.

And cities are beginning to awaken to the notion of call centers are job creation engines (except for New York — the city that never met a tax or burdensome regulation it didn't like). "Enterprise zones" are areas that are tax-abated or otherwise incentivized to allow businesses to open at lower cost, provided they hire a certain number of people.

While this is a good thing, and is starting to show results, people looking at urban locations for their centers have to balance these zoned advantages against higher urban telecom costs, poor services and infrastructure, and possibly deficient educational systems.

Another way to think about your real estate selection is from a facilities point of view: is it somewhere that people will want to work? You're trying, after all, to hire and retain hundreds of people, keeping turnover low. Therefore, safety and security are issues. And along with this go amenities that cost relatively little, but attract workers who will stay and keep them from burning out: accessible day care, parking, even windows.

Telecom. Take it as a given that you need more than you think you do. Imagine the fanciest, most complicated and bandwidth-hungry application you can possibly conceive of using, then double or triple it. If a particular location doesn't make this possible, walk away. Someplace else will make it happen. The last thing you want to have happen is that you commit to a locale that keeps you from using something you'll come to need in three or four years.

Within the US, sophisticated telecommunications is the norm. While it is important for your call center site to have sophisticated telecommunications, you can find this level of sophistication even in the hinterlands.

You might have a particular relationship with an RBOC or a long distance carrier. There may be particular services that you need that you can only get within a prescribed geographic area. For example, centers with extremely high call volumes may find the fact that Kansas City, Missouri is a major switching center for both Sprint (which is headquartered there) and AT&T a big plus.

Government (and other) incentives. There's nothing like getting something subsidized to make a location more attractive. There are lots of ways that governments try to sweeten the pot:

· Real estate tax breaks. They may give you a percentage off the cost of running your center for a few years into your operation, particularly if you're considering a specialized office park that's had government input, or into an economically depressed area.

· Job creation credits. These are incentives based on the number of people you hire and keep on the payroll for a designated period of time. (This is in the authority's interest because they make up the loss in the personal and sales taxes that the hired workers end up paying.)

· Along with that come training subsidies. To get you to locate in an area with lots of underemployed workers, they may offer to pick up the cost of training. This is a good incentive — ask for it if it's not immediately on the table.

· Coordinated sweeteners. That is, a state (or country) can couple one or more tax advantages with favorable rates provided by the local telco, and/or something delivered at the local level by the town or county authorities.

The area of government incentives is one of the most creative (in legalistic and accounting terms) — there's no reason not to ask for what you want and see if it can be delivered, because call centers are so advantageous to a community that there's often no reason not to make them happy. But remember that leverage is a function of size; the more people you might hire, the redder the carpet that's rolled out for you.

A company that's bringing jobs and high-tech facilities into a city is worth incentives. If two or more locations are competing on an even playing field with regard to telecom, labor and amenities, then incentives are a critical lure — and something a call center planner should insist on.

■ International Marketing

Customers in Canada and Europe are just as likely to call you for service as their American counterparts. In fact, some sectors of American industry — travel reservations and high tech, for example — have been setting up call centers outside the US for some time.

Hardware and software companies like Intel and Microsoft have globe-girdling linked centers that answer calls related to the same product lines they sell domestically. Hotel chains and airlines are in the same predicament — they deal with customers who could be located anywhere, and who want to travel into and out of the US.

For companies like those, borders mean little when a customer calls. Other issues come to the fore:

· Answering the call in the right language.

· Taking orders in the right currency — and not losing time or money exchanging that currency back into dollars.

· Appearing transparent to the caller — "non-national," or as little like an American company as possible.

If you're getting the impression that most international call centers are inbound, you're right. For a variety of reasons, telephone selling is not as popular outside the US as it is here. In Europe, that has a lot to do with privacy regulations and restrictions on the way companies can sell over the phone. Lack of customer lists and databases also plays a role.

Other factors include the price of long distance service and incredible number of languages and cultures crammed into one very small continent. All the changes that must be made to accommodate those languages and cultures means that telemarketing does not have the same economy of scale in Europe that it has in the US.

Many companies take their first step overseas with a customer support center. They find that to grow overseas, they need to provide assistance to existing customers already gathered by overseas subsidiaries. These tend to be larger companies.

Look for the same things you would in the US: labor pool, telecom infrastructure, regulations and taxes, education. But look harder, and deeper.

Comparing the incentive packages from different countries (with different currencies, tax rates, languages and levels of technology) can be frustrating.

■ Locations That Want You

United States. Where there used to be definite areas within the US that were call center-friendly, these days there's no real reason why one area is better than another. For sheer volume, the south and southwest seem to be attracting the most centers.

Cities like Las Vegas and Phoenix have surged in population, and in the number of new call centers taking up residence there (or in the surrounding suburbs). There are few good studies that show where call centers actually are; instead, they tend to show where centers could be, focusing on relative cost between regions. So we tend to rely on anecdotal evidence to tell us which areas are more "popular." It's important, then, to keep in mind that most companies that run call centers don't call a lot of attention to their centers. The industry has an imprecise sense of where the centers are clustered.

There are clusters in the Midwest and upper tier of states, through Iowa and Nebraska, South Dakota and Kansas. The states in this region have historically worked hard to attract centers with incentives and on-the-ground assistance getting facilities open and running.

The move to the sunbelt is a more recent phenomenon, driven by lower costs for real estate and (we think) the same thing that keeps people moving into those regions — good quality of life issues, lower cost of living than in the cities on the coasts, lower taxes and the perception of more business-friendly authorities.

Within the US, though, there is a greater diffusion of centers than ever before. The factors that go into call center site selection now include one that was relegated to the bottom of the list ten years ago: corporate convenience. With telecom plentiful, labor expensive and real estate a non-issue, you can put your center near an important client, a critical supply warehouse, a distribution point — essentially anywhere that business conditions dictate.

It's important to note, as well, that the Canadian call center business is not just an offshoot of the American industry, even though most people think of it as a single, unified market. Canada hosts a strong, growing industry in its own right, capable of serving both the domestic Canadian market and the cross-border North American market. Also, you can take advantage of a completely open border, a citizenry with a high education level and a multitude of language skills.

Going "Offshore." Call center agent outsourcing in Europe, the Middle East and Africa (EMEA) will grow significantly over the next five years, as companies look to outsource non-core competencies — that, according to a report from analysts at Datamonitor.

According to their report, Call Center Outsourcing in EMEA, of the nearly 1.2 million agent positions in the region, 150,000 (12%) are currently outsourced to a third party within EMEA.

By 2007, Datamonitor predicts this number will have almost doubled to 290,000 — 16% of all agent positions in the region. Regarding offshore agent outsourcing, the report finds that despite companies remaining averse to risk, the number of offshore-outsourced agent positions will reach 60,000 by 2007. Regions fast becoming popular with companies for offshore agent outsourcing include North Africa, South Africa, Eastern Europe and India and South America. The number of offshore-outsourced agent positions in the Middle East and Africa will grow at a compound annual growth rate of 22% between 2002-2007, significantly above the EMEA average of 14%. By 2007 there will be 40,000 offshore-outsourced agent positions in the region.

For companies looking to significantly reduce call center costs, moving to a location with lower labor costs is becoming increasingly popular, and there are three particular regions in EMEA. For French companies, North Africa, primarily Morocco and Tunisia, are popular locations, while South Africa is popular amongst English and Dutch companies.

The other large growth market for offshore outsourcing within EMEA is Eastern Europe, in particular the Czech Republic, Poland and Hungary. By 2007, there will be 20,000 offshore-outsourced agent positions in Eastern Europe, mostly serving Germany and Austria.

For British and Spanish companies, the main offshore markets are India and South America respectively. The report finds that there will be a shift to these locations, but that the growth in domestic outsourcing markets will continue to drive growth in these countries.

The tendency to look to outsourcing varies greatly by vertical market. Companies in the telecom and technology sectors have a very high propensity to outsource. Pressures to cut costs in these industries will lead companies to outsource even more over the next five years.

Financial services companies are relatively conservative when it comes to outsourcing, but the size of the financial services call centers market makes it also one of the largest outsourcing market.

"More and more companies are looking to outsource their call center operations as they realize that in most cases they are not core to their business." Says Datamonitor's Robin Goad. "Offshore outsourcing is growing rapidly, but from a relatively small base.

Although the savings can be significant, the perceived risks are too many for the majority of companies and offshore outsourcing remains a minority interest. Companies can still make significant savings and productivity improvements by outsourcing to a domestic provider. What we are seeing is the emergence of a combines outsourcing model, where companies keep high value call center activities in-house or in the same country, but are increasingly willing to move low-value and labor intensive traffic offshore

If Europe's your goal, the first question an American firm should ask itself is this: do I need a pan-European center, or do I want to target my center to a single country's market?

A pan-European center requires you to staff up for calls in a multitude of languages. You'll need switches and software that can handle skill-based routing, and probably a voice processing system to offer language-based prompts.

Another option is to identify the call's originating country using the phone network itself. The advantage is that it's more transparent to the caller if your rep answers right away in the language of the caller, without voice system intervention.

What are your options in Europe? You have several really good ones. The top-tier countries are already so similar in the quality of their telecom, that they are differentiating themselves on the basis of pricing, incentives, multilingualism and local regulations that help (or hinder) call center activity.

For a while, Europe was divided into two camps: those countries that had an active effort to attract call centers, and those that couldn't see the benefits and did nothing. Ireland, the Netherlands, the UK and Belgium formed the core of the first group, France and Germany the second. Now both France and Germany have awakened to the facts of life, and are trying hard to attract call centers, particularly those that serve the lucrative pan-European market.

For its part, the UK has prospered because it has a strong domestic call center industry of its own — as many as 3,000 to 5,000 call centers focusing on British customers, by some counts.

The Americas. One of the great open-ended questions of the day regarding call centers is how deeply the North American industry model will penetrate into other emerging markets, particularly Latin America.

A report from Datamonitor on call center location trends suggests that Mexico has emerged as a stable and financially practical alternative for call center investment. With staggering growth forecasts indicating that the number of agent positions will rise from nearly 51,000 in 2002 to over 190,000 in 2007, Mexico is one of a select number of countries that has successfully established itself as a viable offshore locale for servicing Spanish-speaking customers.

Datamonitor's report says that Mexico has a number of distinct advantages in attracting call center investment. In addition to its political stability, relatively sound economy, and close geographic proximity to the United States and South America, Mexico is a member of the North American Free Trade Agreement, and possesses a skilled, bilingual and inexpensive labor pool.

With a steadily expanding Hispanic population in the U.S., Mexico is able to leverage its qualified and inexpensive labor pool to service these customers. The U.S. government estimates that in 2000 alone, over 20 million Americans were Hispanic, representing 12% of the country's population.

While Mexico represents a sizeable portion of the total call center market in CALA, with over 200,000 agent positions across the Caribbean and Latin America, there are a number of additional opportunities to exploit. By 2007, the number of agent positions will climb to over 670,000 and while overall growth for the region is promising, an array of challenges facing some of these nations may stifle near-term success.

The region has been racked with political instability, mounting foreign and domestic debt, and deflating currencies that increase borrowing costs, which in turn, fuel investor uncertainty and exacerbate economic difficulties.

Datamonitor's report on the region says that while countries in economic turmoil such as Argentina will not witness exorbitant call center growth in the short term, others have demonstrated remarkable resilience in the face of Latin America's domestic and international woes — chief among them Mexico and Chile.

In one interesting twist on the offshoring model, one company is using IP technology in a key role in contact center outsourcing. The outsourcer, Orbis Telecom, has implemented an "ACD over satellite" solution with phone lines terminating to their CosmoCom ACD in Florida, with an IP connection to agents sitting in Ecuador, serving the Spanish speaking population in the US.

In Ecuador, all that's needed are commodity PCs with headsets — no proprietary technology to manage at the remote location. Orbis simply manages the desktops.

Orbis's multi-site contact center is based in Florida and includes an offshore facility in Ecuador with bilingual Spanish/English-speaking agents.

Orbis provides both inbound and predictive dial outbound services in Spanish and in English. Knowledge of both languages is especially important for outbound services, since the preferred language of the person answering the call could be either one. With the all-IP CosmoCall Universe, Orbis reduces the cost of supporting Ecuador-based agents by using VoIP (Voice over IP) via an innovative satellite communication link. Skills-based routing ensures that calls always reach the most qualified agents available.

"Contact center outsourcers are looking for ways to leverage lower cost offshore human resources while maintaining quality service. But unifying onshore and offshore facilities with traditional telephone circuits can be a difficult and costly adventure," said CosmoCom's CEO Ari Sonesh.

India. India is developing one of the most robust technical infrastructures for call center operations in the world. In our conversations with Indian technologists and call center managers, we are hearing a sense of confidence that, in general, they are founding their new call center industry on advanced technology (moving swiftly to create Web/voice combo centers, for example). The nation also has the advantage of a highly educated, English-speaking, tech-savvy workforce at a lower wage rate than in Europe or the US. This has exploded into the industry's consciousness.

QUICK TIPS

1. When meeting with representatives of an attractive call center location (government and telco officials, particularly), ask them to bring local educational professionals into the conversation. See if they can coordinate the creation of a call center certification program at a local community college or vocational institute, or at least highlight existing courses that can be used to channel workers towards call center work.

2. A good drawing card for some communities is a nearby military base. Spouses and children of military personnel can flesh out an educated full- or part-time workforce. And a locality with a military presence (or even a former presence) is likely to have good infrastructure.

3. Don't rely on the location authorities to ply you with studies showing why their location is cheapest — too often they will compare their location's costs to some expensive place you never planned to consider, like New York. This is one case where it pays to get a consultant, someone independent who can analyze the differences between places you've chosen to compare.

4. Don't expect outbound telemarketing to be a big hit in Europe. Regulations vary by country, and can be very strict about things like when you call, who you call, and most important: what information you collect about the people you call. If you plan on a pan-European center for outbound, work the regulations very carefully before committing.

5. Make sure that any site you select has multiple points of entry for your telecom — at least two ways into the building from the central office or point of presence. This is in case of accidental failure.

6. Here's something to think about before you commit to anything on the ground — to what extent do you plan on incorporating remote agents (or "telecommuters") into your scheme? Because if you plan on doing a lot of this, you may not need quite as much facility as you think you do. Also, you may not need to plan for the same kind of growth curve that you would have needed five or six years ago. For one thing, there's a lot of call center capacity out there, and outsourcing services are readily available. For another, there's a lot more you can do with smaller, "lighter" centers that are connected to each other, or to workers at home.

India has a large (250 million) educated and highly available middle class labor pool and their compensation costs are only 1/10th to 1/8th that of similarly skilled agents in the US and the UK. The large middle class is creating a growing domestic market for call center services.

High quality labor availability there means very low turnover — no more than 6-10%, , saving companies who open in India hundreds of thousands of dollars in staffing and training costs alone. The country's large population will stave off the wage and turnover cost pressure experienced in the US for some years to come.

On the flipside, however, one British report pointed out that though in-house call center executives clearly saw the labor cost advantages of outsourcing to Indian firms, they were concerned about accent, power and communications continuity and the country's political and fiscal stability.

When examined by vertical market the results showed that the IT/software, telecom and other outsourcers sectors were more inclined to consider outsourcing to India.

In short, when you're undergoing a site selection process, open your eyes to all the new possibilities that are out there, domestic and foreign. You should look off the beaten path. There are bargains to found, and attractive locations that can serve very well as the long term home of a call center. The important thing is to be flexible, and work with the community economic development officials. You don't need to be in a big city, or in the Midwest, or even in the US. It's all up to you.

Chapter Three

Facilities & Design

The furnished environment your agents have to work in for four to eight hours a day affects their attitude more than whether the technology they are using shaves a few seconds off call duration. Their comfort, or discomfort, within that environment has an undeniable effect on the way they deal with customers. And of course, on turnover, which affects long term hiring and training costs.

But there's more to call center design than picking out pretty colors and sleek workstations. The right call center furnishings can help the work get done faster and better. Employees are happier, they are out sick less, they sell more and serve your customers better. Here are some of the important factors that go into a successful design.

One thing we're not going to get into here will be networking and cabling. This is a thorny issue, and one that changes as frequently as any other technology outlined in this book. Suffice it to say, make sure of these three things:

1. You choose workstation units that allow easy access to cabling and phone wires.

2. Your architect and design team are fully aware of the kind of voice and data networking that you want to install, and that they are sensitive to the pecu-

liar needs of expensive telecom and computer equipment. (In other words, don't put the ACD in a room next to the HVAC system or the cafeteria).

3. Whatever your wiring plant, make sure it's upgradable without having to rip apart floors, walls and ceilings to do it.

It's also been argued by some architects and designers that the physical layout of the call center can have a direct effect on a company's profitability.

One study done in 1990 of 70 million square feet of office space found that the cost to build and maintain office facilities is only 15% of the total, with the remaining 85% going to salaries. The argument would then go that if you focus the design to have as beneficial an effect as possible on the workforce that comprises that 85% outlay, you can make the business case to spend more up front on layout and design with the expectation of back-end savings. Substantial research has been completed that demonstrates the effect design can have on increased productivity of building occupants.

Times have changed. In the past you could say the chairs, the lighting and the design of the workstation were the most important elements of call center design. Those things are still important, but issues of health and safety are increasingly on people's minds these days. Yes — even in call centers.

One architectural firm we spoke to ballparked the cost of fitting out office space for call center use as ranging from $22 to $38 per square foot. Assuming the average cost of $30 per square foot is borrowed at 9.5% for 10 years, an increase in the staff productivity of 15% will pay the debt service for the entire construction cost necessary to improve the space for call center use, they say.

▬ What Affects Design?

The ADA. The Americans With Disabilities Act (ADA) is a federal law that can affect call center design. Most important, the ADA is civil rights legislation, not a building code. You can't get around it by moving from one location to another. It's also very vague, forcing you to go out of your way to respond to whatever circumstances present themselves.

Some states have building codes that require ramps, accessible restrooms and other accommodations for the handicapped. The ADA does not require you to have any of these things — but it does make it illegal to reject a qualified applicant because your facilities are not accessible to him or her.

That means that you have to plan ahead. You have three choices. You can put these facilities in at the beginning, when it's going to be the cheapest and least disruptive. You can put them in when you find you need them, that is, when you hire someone who needs them. Or, you can ignore them completely, and face the lawsuit.

Sometimes teleservices firms find out too late that their choice of workstation doesn't accommodate a wheelchair, or a random floor plan makes it difficult for a blind TSR to get around. Good call center design recognizes the relationship between people and the physical constraints of the workplace.

Safe work environment. Human safety should be your first priority. Make sure everything is fire resistant, that the exits are appropriately open and marked, and that all fire safety regulations are followed. And that call center personnel are aware of fire safety procedures.

This is so basic to facilities design I shouldn't have to mention it. But I will anyway.

The call center's application. For order entry applications, for example, one expert recommends at least 35 square feet for each person's workspace. But for customer service, you might need more, up to 45 square feet. That's because service and support reps often have to refer to manuals, documents, and other peripheral materials that should be stored within easy reach.

You'll also want to account for the number of people in groups or teams, and the position of supervisors and team leaders. You're going to need room for meetings, for example. And you're going to need semi-private call center stations that can be isolated during training or coaching sessions.

Corporate cultures may dictate particular placement, and that too should be recognized when planning the layout.

The relationship to other departments. Remember, once the center is active, you're going to be watching the length of calls as a key component of costs and productivity. If call center agents are running up and down halls to another part of the building regularly, you didn't plan well. If they need to be in constant contact with the fulfillment department, for example, work that out ahead of time.

If you can't physically bring the two departments any closer, then explore some kind of automated networking solution that will tie people and systems together — that may alleviate some of the distance trouble.

What support systems do you need? Will you need a cafeteria? Consider long term plans and the company-wide flow of traffic. How many conference rooms will you need? Where is the copy room, the time clock? Will you have central or local filing?

All these questions must be answered in advance. But it's critical that these questions be answered by call center management as well as by the architects and the company's upper management.

■ The Existing Center

Those ideas are fine, if you're building a center from the ground up. But what if, as is more likely, you're rehabbing a center that's been around for a while, or outfitting an expansion of a center?

Luckily, there are still plenty of facilities factors under your control.

Lighting. Indirect lighting is the best if you can afford it. If not you should use florescent pink tubes and parabolic lenses. These lenses diffuse light straight down to eliminate glare. Full spectrum fluorescent tubes are available from some manufacturers that give a natural sunlight-like illumination.

Full-spectrum lighting is color balanced so there's no yellow tint and less glare than with florescent lighting. The tubes fit into existing fluorescent fixtures.

Noise. Nothing is noisier than a roomful of people all talking at once. It's hard on the employees, and it makes callers think they're calling a roomful of people. There's nothing so unprofessional as a call center that sounds like a.... call center.

If you want the both the caller and the rep to feel more comfortable, try acoustic wall paneling, and if funds allow, white noise machines to diffuse noise. Using sound-absorbing foam or tiles on the ceiling, walls and other soft surfaces, and carpeting, keeps the sound from bouncing around. Plants are also good for the air and absorbing acoustics, but that's a minor fix at best.

What some centers use are the same kind of foam tiling found in recording studios, though this can give a closed-in look to the place. In a cubicle environment (which most call centers are), talk to the manufacturers of the workstation units themselves about what kind of acoustical absorption properties they build into the wall coverings.

And of course, noise-cancelling microphones in the headsets will help keep the apparent volume down, though that's not strictly speaking a facilities question.

Seating. Your full-time agents spend at least seven hours a day at their cubicles sitting. The chairs you choose mean a lot. A chair affects posture, circulation and pressure on the spine.

We recommend chairs with height-adjustable armrests, split backs that hug your back (to relieve pressure on the spinal column), a moveable seat and an adjustable back angle.

We're not saying you have to go out and buy everyone a $1,000 Herman Miller chair, but don't put them in a $39 OfficeMax special, either. That's putting yourself on the fast train to a high turnover rate.

Monitors. The top of the screen should be at eye height or slightly below and about 18 to 24 inches from the eyes (30 inches if you are concerned with electromagnetic radiation and your monitor is unshielded). The monitor should swivel to help reduce reflections. Once again, this is a small thing, with a really minimal added cost. But buy them the biggest monitor you can, especially if they're going to be looking at a screen that pops a lot of critical customer information into a lot of tiny windows. The larger the monitor, the larger you can make the type in all those tiny windows. Seventeen inches ought to be the minimum.

Wall height. High walls between employees reduce noise, but they also cut agents off from one another and reduce collaboration. Sometimes the best way to deal with a call is to lean over the partition and ask another agent.

In the past, it was also important that agents be able to see a centrally hung readerboard. Now, with scrolling screen tickers full of ACD info, you're not so dependent on that, so you can consider not only higher walls but a less formal cluster organization of the cubicles. One generally accepted height is 42 inches. That gives a certain amount of privacy without shutting the agents off from what's around them.

Agent input. Agents ought to have some say in how call centers are designed. They're not the only ones who benefit when you give them input — managers and supervisors get happier, more productive employees and fewer compensation claims.

Today, more and more call centers are collecting input from their employees before buying workstations, for the simple reason that they want to keep those employees as long as possible. Because call center agents must perform repetitive phone and keyboard tasks and spend all day (excluding breaks and lunch) at their desks, using ergonomic equipment is crucial. You'll get happier, healthier and more productive employees. In the long run you'll save a bundle in time and money since you'll have lower turnover and better morale.

Considering there are more employees suing now than ever before for repetitive stress injuries (reported incidents of RSIs are higher than ever, accounting for 60% of all occupational illnesses) there's no better time to offer courses in prevention and re-evaluate your center's set-up.

The workstations. There are a lot of options in buying and coordinating the placement of the actual seats where agents will do their work. We're not talking just about cubicles here; call center workspaces are carefully designed and constructed for the particular needs of this industry by a number of specialty companies.

This kind of thing is often overlooked, or put aside as managers think more about the critical (and expensive) technology and hardware they need. It's easy to forget that labor is the single biggest ongoing expense in a call center. Intelligent workstation design is an easy way to reduce costs over the long term by keeping turnover low and employees happy. The type of workstations you choose can facilitate team building or discourage it.

There are three types of workstations: the cluster, a pinwheel like setup with four to six work areas sprouting from the core in the middle; the rectilinear, a traditional panel system with four wall panels at each station set up in rows; and the modular or free-standing workstation.

One vendor says that cluster workstations are beneficial to companies, like large catalog or insurance companies, that need to put many telephone- and computer-intensive workers in the same room. That's because clusters let you fit more people into less space, but the people don't feel cramped.

In fact, the cluster arrangement lets you save 10% to 25% of your floorspace and doesn't give you that mousetrap/maze effect that rectilinear workstations sometimes create.

The gentler floor plan makes it easier for people to walk through the call center and between groups, fostering teamwork.

One downside to the circular workstation arrangement is that the partitions between stations are sometimes too high, making communication between agents on the telephone difficult. With the cluster it's easier to talk between workstations, but hard for people to come in.

Rectilinear, or panel, workstations are a good choice for centers that need more space for each agent or that need more flexibility in panel and desk heights. The design of a center around these stations is more forgiving, and easier to change as conditions change. The work surface can be moved between notches in the side panel to accommodate wheelchair-bound agents or agents of different heights. Rectilinear workstations are popular choices for engineers, managers, people who need extra room for storage cabinets and anyone who has conferences with coworkers. You also find this style the preferred on in technical support centers, where the reps have to refer to a lot of external materials — binders, reference manuals, and so forth. The type of workstation you choose should complement your company's team-building style.

When evaluating workstations:

· Look for a style that's easy to install and reconfigure. Look for something that doesn't have too many parts and pieces, but where you can add overhead shelves and in/out boxes.

· Make sure the equipment can be connected within a panel, rather than to a box that sits on top of a desk.

· Buy through a local dealer so you'll have nearby on-going support. And a dealer can help with things like placement of workgroups for departments who need to communicate regularly.

· Look at it as a strategic investment. Chances are, you'll have it for the next ten years, so you don't just want to look at price. It should be pleasant and functional.

· Get panels with metal frames because they're more durable than wooden ones. Also, get fabric panels that can be re-covered if damaged. And again, examine the acoustical properties of those panels.

QUICK TIPS

1. Is your building sick? Sick buildings are ones in which problems with the structure cause people physical distress: problems such as air quality, comfort, noise, lighting, and ergonomic stressers (poorly designed workstations and tasks). Here are some questions to ask: Is there a source of contamination or discomfort indoors, outdoors or within the mechanical systems in the building? Can the HVAC system control existing contaminants and ensure thermal comfort? Is the HVAC system properly maintained and operated? And, do the building occupants understand that their activities affect the air quality?

2. Make sure your architectural and design teams are fully aware of the kind of voice and data networking that you want to install, and that they are sensitive to the peculiar needs of expensive telecom and computer equipment. For example, you don't want them putting the ACD next to something that's going to be vibrating, like an HVAC system. Believe it or not, those considerations are sometimes overlooked.

3. If you're putting in a brand new center in an area that has (or might have, in the future) a tight labor market, consider the amenities that will attract agents. Things like good parking, broad windows with a landscaped view, onsite day care (this is a good one). Anything that keeps agents from leaving is a good thing.

4. Keep the center clean! There's nothing so depressing as spending millions on a facility, and then having it degrade within months because people can't keep their workplace clean. Good habits start at the top: the company should hire a cleaning service to maintain the site.

5. What will you do about smokers? Granted, many places have gone completely non-smoking indoors, and that's good. But you don't want to cut yourself off from the labor pool that includes smokers. Nor do you want smokers to sit at their desks increasingly irritated, because that's going to show up in customer interactions. Find a place they can go, outdoors, and make it a safe and comfortable place, close to the center's floor (so they're not spending all day travelling to and from their desks to the smoking area).

6. Don't forget security. If you're running multiple shifts, you'll have a hard time getting people to work there in the evenings and on weekends if you're not providing a safe environment. And that safe zone should carry outside to the parking lot and the place where the smokers congregate, too.

7. When designing your center, don't neglect large, common, non-work areas to be used for training, motivational sessions, staff meetings and the like.

■ Future Design

As call centers focus more intensely on retaining their best agents and reducing turnover, human issues will come more to the fore in designing (and redesigning) call centers.

Centers will include more "community" areas: conference rooms, training centers, even classrooms. And as call centers begin to respond to more than just traditional voice phone calls, expanding into areas like email and Web response, as well as possible video calling, the kinds of workspaces that will be required will undoubtedly change. The smart design team will take these things into account now, because a call center is a five to 20 year commitment.

The best way to ensure that a call center is ready to accommodate your needs in 2010 is to allow for substantially more layout flexibility than is typical today.

Components that should be considered include:

· Uniform, ceiling mounted indirect lighting systems that are layout-independent.

· Furniture literally on wheels or furniture systems that can be reconfigured overnight.

· Raised floors that allow ultimate cabling flexibility.

That way, the call center will be completely adaptable to any changing business circumstance, whether it's driven by new technology, new ways of operating, or changing company cultures and ideas.

Part Two

Routing Calls: Switches & Hardware Systems

The ACD is the heart and soul of the modern call center. It is the engine of productivity — the single piece of technology without which the whole edifice of inbound sales, order taking and customer service all crumble.

Chapter Four

Toll Free & Long Distance Services

After the center's physical parameters are set, and the agents are hired, the most important element (at least from an ongoing cost standpoint) is the pipeline into the center. The toll free and long distance services that you choose will be so expensive, and yet so rich with features and possibilities, that it's imperative that you choose carefully, and that you revisit your decision again and again for as long as the center operates.

Toll free service was once amazingly simple. You had one company to buy from, and very little leverage in the kinds of pricing plans and service offerings you could get. By very little, I mean: none. Companies didn't begin to build call centers until there was a cost-effective means of making nationwide toll free calls, roughly thirty-some-odd years ago.

Wide Area Telephone Service, originally an AT&T creation, was the first iteration of toll free. It discounted long distance service, put the cost onus on the called party, and so began our journey down the call center road.

With divestiture and long distance competition, there were naturally more choices and the price of call center telecom began slowly to descend. And then, in the early 1990s, just as the three main long distance companies competed fiercely in a very public battle for the home consumer long distance market, a not-so-public but just as vicious fight for the call center market heated up, too. It was helped along by resellers and aggregators, which are essentially secondary marketers of long distance service. Resellers would buy bulk minutes from phone companies at a tremendous discount, and resell them at a very small profit margin, making money on the spread. Aggregators would combine the telecom traffic generated by lots of small companies until they were able to go to a phone company and commit to buy big packages of minutes, hence qualifying for the same deep discounts the telcos gave to their largest customers.

All these things worked to drive the cost of a long distance or toll free minute down past 10 cents, in some cases to as low as five. Of course, things are never as simple as they seem.

You almost never buy telecom minutes just bare — they are just the beginning of the process. It's all in the value-add. What's a long distance package without some kind of service assurance policy, for example, or without network reliability guarantees?

Or better yet, would you pay more per minute if the carrier let you manipulate the network according to your own traffic needs? Routing calls here for one reason, there for another — that's a pretty powerful ability and they all have it.

What about being able to hold calls in the network, instead of queuing them up in your ACD? Or park them in the carrier network, while the net queries the ACDs at several centers to determine which one has the right person to answer the call? You can do that too. The more complex the routing dynamic, the more likely it is that you'll have to go to someone other than the carrier for the actual software that makes it work, but the carriers are now eager to help hook you up. (They were not always so eager; phone companies tend to be less than far-sighted, as technology companies go.)

The carriers have also experimented, with mixed results, with services that actually perform transaction processing, even fax processing, in the network. It's like having an outsourcer handle your calls and your transactions, but there's no actual outsourced center; it all happens automatically.

They want you to take advantage of a lot of these advanced services, whether or not they provide the mechanics, because quite frankly, call centers are a gigantic consumer of telecom minutes. The more time your callers spend hanging out in their networks, the better off they are.

Add to that one other critical reason. If you posit the notion that long distance and toll free are pretty much the same from carrier to carrier, that all are equally reliable, clear, inexpensive and available, then what keeps you from hopping from one to another at the drop of a hat? They hook you by getting you to buy ancillary services. I can foresee a day when the value-added services are more important to the carriers than the presentation of transmission minutes, and they end up giving the minutes away to their best customers as a loss leader. Especially when we enter a world with packetized networks and all sorts of alternative transmission methods that reduce the actual cost of moving a call from here to there to effectively zero.

So what was once simple — buy on price — has become complicated. But wait, there's complexity on another level. Until 1993, if there was a particular phone number that you wanted to have in the 800 toll free code, you had to buy service from the carrier that had custody of that number. You had no freedom to change carriers and bring your number with you — if you had significant brand equity built into your number (800-CAR-RENT, for example, or 800-MATTRESS), you were stuck.

Until 1993. That was the year that *800 Portability* reorganized the way 800 numbers were given out, and changed the whole dynamic of how you acquire and route 800 numbers. Portability meant (and still means) that you have custody of your toll free number. You can keep it if you want to change carriers. This, of course, gives the carriers added incentive to serve you better, to offer more interesting features in their toll free networks to keep you as a customer, now that you're not a hostage.

Remember also: they need your business. Call centers are monster consumers of toll free and long distance service. They will make deals with you. If they do not serve you well, you can and should leave. In fact, you should absolutely have arrangements with at least two out of the three main carriers for your core service. At a minimum, that protects you against service outages. But it also allows you to compare, month by month, the offerings and prices they charge.

At first there was a lot of concern (generated by AT&T, in part) that portability would cause degradation of service (especially longer call set up times) because each call to a toll free number has to be passed along a more complicated pathway to query a database and determine which carrier routes it before it can be connected. Happily, those problems never materialized. Portability became part of the competitive landscape, and I think was a strong factor in the rush to grab 800 numbers a few years back. That rush, in turn caused the 800 number series to run out and forced the opening of first 888, and then long before anyone thought possible, another series, 877. (Other reserved series are warming up in the bullpen.)

Portability made toll free an intelligent network application. Users with multi-site centers who wanted features like Least Cost Routing, or sophisticated queuing options benefited immensely. Many of these services are expensive, though. In some cases they can add as much as 50% to the cost of a call, putting the options out of reach of many small and medium sized call centers. High volume users have been the main beneficiaries of price-cutting and volume discounts, leaving smaller users with higher costs and no appreciable gain in service.

Through bundled consulting plans and alliances with hardware manufacturers, the three majors are trying to be more to you than just a series of trunks and switches. Offering everything from complete outsourcing of your center to simple "press one for" service, phone carriers are providing more options for call centers than ever before.

Will the call center of the future be paying for carrier services by the transaction instead of by the minute? This is just one of the possibilities raised by the brave new world of call center offerings from the three major long distance carriers.

In previous editions of this handbook, I detailed precise product offerings from the main carriers. I'm not going to do that this time out, and here's why. Information like that changes so rapidly that there's no guarantee that they won't have changed a brand name, feature set or pricing schedule by the time this book gets into your hands.

Instead, what you'll find here are broad brush outlines of what kinds of things you can do with their services. Please don't beat me up if you try to buy these services from the carriers and they've changed some from the way they're presented here. For more details about exactly what they do offer, I recommend visiting their websites. In general here, I'm going to refer to these services as toll free, because that's what call centers are most interested in. This isn't a bias against outbound-based centers, it's just a fact that inbound call routing is more complicated, a more feature-rich set of tools, and because it's more expensive, you have to work harder to get exactly the right deal.

AT&T. AT&T has the most to lose in any competition for long distance or toll free services. By ridding themselves of Lucent, they said to the world that what they wanted to be was a transmission company, once again a true carrier rather than a phone systems company. And they did that after seeing exactly how cutthroat and expensive a battle for market share can be. So they must really mean it.

One of the more interesting things they've come out with is called *Transfer Connect with Data Forwarding*. Data Forwarding uses ISDN technology along with computer telephony integration to let your agents forward each customer's data along with their calls. It lets the agent or voice response system that initially receives the call forward the customer information to the receiving agent. This information can range from name, address and account number, or application-specific data like frequent flier information, insurance plan specifics or personal IDs. The data instantly pops up on the receiving rep's desktop.

There are packages that let you route by time of day, geographic origin of the call (down to the exchange), and countless custom preferences that you set and reset any time you like.

Next Agent Available Routing lets you reroute toll free calls to up to 99 alternate locations if the primary location is "busy." If the primary location is busy, calls are instantly rerouted to the first available termination, decreasing customer "on-hold" waiting time. You can customize NAAR to match your call volume needs: you determine what is considered a "busy" location by defining the Maximum Calls Allowed (MCA). You can set this value in advance and override it at any time. When the MCA threshold is met, all toll free calls will be automatically rerouted to an available termination.

With *Network Queuing*, you can automatically queue calls without investing in any additional equipment. This feature allows your call center to optimize call distribution and improve call completion rates. Calls can be queued for multiple call centers or for locations with a single queue. Network Queuing can help boost sales by preventing customer hang-ups and increasing call completion rates. Of course, you can do this with premise technology built off the ACD, so you have to plan in advance where you want your call control to be handled.

Sprint. Sprint's basic toll free offerings are designed for companies whose calls terminate at one location. They are pretty basic, but they'll suit the vast majority of single-site small- and mid-sized call centers. You get things like DNIS and ANI call identification popped to the agent screen. (A simple thing that can shorten calls by 10 to 20 seconds, and when you add that up, call after call, multiplied by dozens or hundreds of agents, that's a lot of money.)

You can distribute calls across a trunk group (very rudimentary), and designate a secondary location for calls to terminate in case of overflow. Sprint

QUICK TIPS

1. Use multiple carriers. That way you don't face catastrophe if one of the carrier networks suffers an outage or overload. Try a percentage allocation; that way, callers might face more busy signals during trouble times, but they will ultimately get through.

2. Hold the carriers to their service assurance plans. And make sure you know exactly what to do to put that plan into place. If calls are supposed to terminate at another location or through another number, make sure the secondary center knows they are part of that plan, and make sure the internal mechanics are in place to handle the load. But it's essential that you know how fast the carriers are supposed to cut you over, and that you make them do it.

3. Squeeze the carriers for everything you can get out of them, including Internet service, and security. The margins for long distance services are so scary that you have to wonder if even the big ones will still be in this business a few years from now. Hardly anyone wants to be a long distance carrier anymore, even for the once-lucrative business services sector (read: call centers). You should be in the driver's seat. Squeeze every drop of value-add you can onto your long distance plan. And check your bills rigorously. They still make a lot of mistakes.

also offers a Carrier Diversity program that helps manage and coordinate service provided by multiple carriers through a single point-of-contact. You can allocate calls to carriers by percentages you set or based on location, time of day, day of week and day of year. (Very sporting of them.)

More advanced service includes what they call Network Call Distributor, a sort of virtual call center facilitator. NCD collects activity information from each ACD in your system every 20 to 60 seconds. It then uses that information to automatically route your toll free calls to the best location at that time. They have another, similar feature that's for Call Allocation, distributing calls to your toll free number across locations. You specify a percentage of the calls for each location, matching your call volume to each location's capabilities.

The Future

Elsewhere in this book you'll find information about the coming together of call centers and the Internet. It's impossible to leave the subject of buying telecom transmission services without saying something about what the Internet revolution will do to call centers.

There are dozens more that provide advanced networking services to large businesses that include call centers. We are poised on the edge of a new networking world, one in which a bewildering array of higher bandwidth networks will be available for piping into your center. Frame relay, ATM, xDSL, ISDN, all sorts of IP permutations — they are strange, and the market economics haven't been worked out yet. I don't know yet what combinations of voice and data traffic will win out, and which carriers will offer what two or five years down the road.

What I do know is that, as the pipe grows thicker and stronger and more varied, the cost of each individual call gets lower and lower. As I said before, it's going to be the value-added services that differentiate carriers. It's going to be harder to tell who is a carrier, as data networking melds with telecom. And it's certain that some form of digitized packet-based traffic will be part of even the most traditional network, driving costs down and opening up new international dialing vistas. The moral: don't box yourself in, get the best deal you can and prepare to turn on a dime when someone offers you something cheaper or throws a killer application on top of their core service.

Chapter Five

The ACD

Automatic Call Distributor. A simple way of describing not just a piece of hardware, but really everything that goes on in a call center. That's the basic function, anyway: taking incoming calls and moving them to the right place, the agent's desk.

Over the years, things have changed. The ACD is responsible for more than just moving calls. That's probably not even the best term for it anymore; I should probably be using something closer to *telephony server*, though that term is already in use for the LAN server that moves call control commands from client workstations to the attached ACD or PBX.

Now, the ACD's job is not just to route calls, but to manage the information associated with those calls as well. "ACD" is really a function that can be carried out by a wide variety of different kinds of processors.

At the very low end, you can buy a PBX that has "ACD" (read: call routing to agents) built in, or you can choose to add it on through a PC application. (Yes, the switches are generally open enough to third-party or other add-on apps.) You can route calls in the network, thanks to intelligent features built into the carrier networks.

The ACD is the heart and soul of the modern call center. It is the engine of productivity — the single piece of technology without which the whole edifice of inbound sales, order taking and customer service all crumble. What the ACD has done is enable the volume of calls you take to escalate intelligently, and in ever more specialized complexity. It has matured beyond call routing. It is the brain and control point for the call center, for both inbound and outbound, for voice calls and data traffic. It's a call center's arbiter: setting priorities, alerting supervisors to patterns and crossed thresholds.

Once the term "ACD" meant a very specific type of telephone switch. It was a switch with highly specialized features and particularly robust call processing capabilities that served at least 100 stations (or extensions). It was purchased mostly by airlines for their reservations centers and large catalogs for their order centers.

Companies with less specialized needs bought different technologies that didn't offer the same specialized features. Currently, true ACD functionality is found in telephone switches that range widely in size and sophistication.

Today, there are PC-based ACDs, key systems with ACD functions, key systems that integrate with a computer and software to create a full-featured ACD, PBXs with ACD functions, PBXs with ACD functions that are so sophisticated they compete with stand-alone ACD systems, stand-alone ACDs that serve centers with less than 30 agents, traditional stand-alone ACDs (don't misunderstand, these are usually the most sophisticated), ACDs that integrate with other call center technologies, and nationwide networks of ACDs that act as a single switch.

There is simply no technology more suited to routing a large number of inbound calls to a large number of people than an ACD. Using an ACD assures your calls are answered as quickly as possible. It can provide special service for special customers.

ACDs are capable of handling calls at a rate and volume far beyond human capabilities, and in fact, beyond the capabilities of other telecom switches. They provide a huge amount of call processing horsepower. Using an ACD assures your human resources are used as effectively as possible. It even lets you create your own definition of effectiveness. An ACD gives you the resources to manage the many parts of your call center, from telephone trunks to agent stations to calls and callers to your agents and staff.

With all these call handling options, many of them surprisingly open and modular, why would anyone still want an expensive standalone ACD for their center? Two simple reasons:

Power. Nothing has more raw call processing ability than a first-tier standalone. Nothing else is so uniquely suited to the needs of today's reservation or financial service megacenters.

Technology. When it comes to integration with other call center systems, like IVR, data warehouses and intranets, nothing can beat a powerhouse ACD. The same goes for multi-site networking and skills-based routing — two of today's most sought after inbound features.

No knock on smaller systems like PC-ACD and PBX/ACD hybrids (which account for much of the industry's phenomenal growth in small centers), but there is no substitute for the call-crunching strength of a standalone ACD in many high-volume applications.

The ACD is being changed by two dramatic trends in the call center. First, it is being asked to channel more information, of many different kinds, in more directions. A decade ago, there were two kinds of information: the call itself, and raw log information about the calls in aggregate. Little else was needed, and if you did need more details about what was coming through the ACD you could analyze the data (which often came out a serial port) on your PC with a cumbersome third-party tool.

Now, call center managers need information — presented in a form that makes it easy to grasp quickly. High-end ACDs vendors have added data management modules at a rapid clip. Also there are many outside programs that can connect to the ACD and funnel data in and out. These include workforce management tools that forecast load, and software systems that put real-time and historical data into any form needed (like reports and readerboard displays).

ACD vendors are improving the tools the supervisor has to tweak the ACD while it's in motion: things like creating groups on the fly, moving calls and personnel around, monitoring for quality, to name just a few things.

The other dynamic change is in what kinds of calls the ACD has to route. Sometimes this is referred to as *alternate methods of call delivery*. Call centers have been integrating ACDs with IVR for years. Now vendors grapple with the Web and the Internet, with calls that come in from PCs and that terminate in databases instead of agents. What's important is the transaction between the customer and the company, not what wire that transaction passed through.

Another thing that's shaken up ACD design is skills-based routing. At first, this was a feature that was added to switches more because the technology was possible and cool than because call centers were clamoring for it. It's taken a long time for call centers to figure out how to make it work because (for reasons that I'll go into in more detail in another chapter) skills-based routing rubs the wrong way against the proper use of workforce management software.

Be that as it may, skills-based routing is a highly interesting and advanced system for distributing calls that come into an ACD. Traditional routing is based on two factors — an equitable distribution of calls among available agents, and the random nature of incoming calls. Skills-based routing changes this somewhat: it routes calls to the agent "best qualified" to handle the call, measuring "qualified" by agent parameters you set.

The ACD does this in two steps. First, some front-end technology must be used to identify the needs of the caller. That's usually accomplished through DNIS, ANI or an IVR system. Then that information is matched against the sets of agent skill groups. There are two ACD advances that let you run skills-based routing effectively:

· Leaving a call in an initial queue while simultaneously and continuously checking other agent groups for agent availability;

· Or allowing an agent to be logged on to more than one agent group (in this case a skill group) at a time, assigning priorities to those groups by skill type.

And finally, changes have been hastened by the need and desire to link call centers together into multi-site call center networks. In some ways, this is operationally an extension of skills-based routing: it's not enough to choose the best available agent — often you have to choose the best available agent at the most appropriate location (based on time of day, traffic at one or more sites, skill clusters or call priority).

To a greater or lesser degree, vendors of standalone ACDs are pushing the technology envelope with their switches. Some are concentrating on software development to add value to the core switch. Others are paying more attention to integration with third-party call center technologies like the Internet and IVR. Still more are adapting their switches to smaller, departmental call centers, hoping to catch some of the growth in the industry that way. In all, it's created a dynamic atmosphere — one in which if you want a feature, all you have to do is ask.

■ Small Center? No Problem

First, let's start by thinking of the needs of the smallest centers, those for whom the purchase of an expensive, standalone ACD is too much to handle. Luckily, call routing is available as part of the PBX configuration, bringing large call center tools down to the level of even the smallest centers.

Think of the five or ten person collections department, the customer service area of a larger company. They have many of the same needs — and problems — as larger centers. But until now, there have been few call handling tools that deliver state of the art features at a reasonable price.

Their personnel are not always dedicated phone reps. They need flexible solutions that build on the systems already in place, that give them room to grow without putting the company in the poorhouse. The response to those needs is a new variety of call handling system — the ACD without the box, or the PC-based ACD. Thanks to the new-found openness of switch vendors, developers are offering a host of software products that add ACD features to key systems and hybrid switches.

For one thing, it's far less expensive to bring ACD features into an existing business phone system. There's no capital expenditure on a big piece of hardware. With larger ACDs, it's very difficult to justify at the six-agent size or, for that matter, anything below thirty agents. It's also a lot more flexible than it used to be. You can easily integrate top-notch systems like interactive voice response or voice mail, giving your small center a highly professional appearance.

Critical to call centers is that you be able to add third-party call control. You don't need to know how to program to set up a rule-based system for getting the right call to the right agent. With the PC it becomes a low-end solution. You can do a lot of things, like provide special treatment to customers based on the language they speak, route calls based on skill sets, or based on time of day for full 24-hour coverage.

What the PBX ACD does is let you dabble in call centers without having to go full bore right away. Make no mistake — one of these low-end ACDs built off a PBX will only get you so far. If you're going to grow beyond a certain point (50 agents is a good ballpark), then explore the larger, standalone systems. At least explore the low-end offerings from those vendors, because those vendors are offering much smoother upgrade paths than ever before, hoping to capture some of the small center market.

PC or PBX ACDs allow small installs (typically 10 to 15 people) to be placed on the same technological plain as bigger centers. Since many companies already have PBXs that can be enhanced with available software, they can dabble. It's possible to convert a few users and then decide that if things go well (they usually do), to expand further.

For example, Cintech offers *Prelude*, by its very name a starter system that encourages people to step up to *Cinphony*, Cintech's more advanced software ACD. *Prelude* is aimed at retail stores, pharmacies, universities, car rental places — places that until now might have gone without features like call categorization and advanced routing.

Comdial's *QuickQ* software lets you set overflow patterns between multiple small groups, and lets you change those parameters quickly, with a minimum of required knowledge.

These systems do not deliver everything you'd expect if you used a dedicated ACD, but they don't have to. Departmental needs are different. Few need multi-site routing, for example. Department heads (who may not be telecom people) need different kinds of reports that have more to do with sales and costs than with call traffic.

There are many small call centers that are just beginning to realize they are call centers. And that they need the same kind of technologies big centers have been using for years. Customers demand the same kind of service, no matter how big you are. This small-scale solution lets them reach more of their potential for pleasing customers at reasonable cost.

■ The Future Beckons

It's a tough time to be a switch vendor. Not too many of the traditional tele-com companies that make ACDs for call centers still want people to think of them as switch vendors. Depending on where they see the future going, they want to position themselves as CRM companies, or Internet compa-nies, or companies that provide one-stop turnkey contact center systems. It's a far cry from the days when the standalone switch was the centerpiece, and everything else you needed for customer contact was hung off of it like balls on a Christmas tree.

What we have now is a modern ACD that is not merely a piece of isolated dumb hardware, but rather an open, network-spanning piece of futuristic technology. The ACD's job is not just to route calls, but to manage the information associated with those calls as well. "ACD" is really a function that can be carried out by a wide variety of different kinds of processors.

Rockwell, Aspect, Avaya, Siemens, Nortel and a few others have fought like dogs to keep up with each other in core technology, with a surprising degree of success. And while the market does segment out into big players and smaller players, into switches mainly for the large center or the smaller cen-ter, the fact is that they've all been pushing each other to stay competitive.

But now selling an ACD is a far more complex business - and discerning the difference between various vendors' offerings is not so simple as going down a feature checklist or RFP.

What's driving the sales of ACDs? According to Rockwell's Terry Murphy, it's no longer the features of a standalone switch. Instead, he argues, at the high end the driver is really the way the technology is aligned with an enter-prise's entire customer care strategy. "This implies the portfolio of value that goes beyond just an ACD," he says, including applications and the net-work capability that surrounds the core switching.

"This is usually driven by the CIO," he adds, pointing out that "for the most part, call center managers have been disempowered."

This is not news - critical analysis of the way the call center and the rest of the company interacts has shown that this trend was developing at least for the last five years, since the Internet enabled the movement of call center data both inside and outside the center. Once other departments had access to call data and married it with other types of corporate information, that data became more important to the company as a whole for purposes that went far beyond the need to measure agent performance and call duration.

Once you start down that road, though, you've ratcheted responsibility for the information up the chain of command - from the call center manager to the head of the MIS department, or the vice president of customer service, or the CIO him- or herself.

What that means for the purchase of ACDs is that the switch has to fit much more tightly into a whole topology of corporate networking and data management. It's not an isolated purchase anymore.

Couple that with the stunning wave of mergers and acquisitions in the past few years, and you really do see a need for switches to operate in a multi-vendor, multi-platform environment.

In Rockwell's case, that meant changing the theory of sales. It means relying more on resellers (Bellsouth, SBC, Norstan and Verizon, to cite the ones Rockwell talks most about). "What those relationships do is bring a portfolio of application and capabilities to bear," Murphy says. In other words, they need to wrap their switching technology in a blanket of all the other things that a more widespread telecom provider can deliver, including the promises of trouble-free integration across open platforms.

The way you end up differentiating one switch from another, therefore, has more to do with issues of scalability and reliability than with specific features. This, because the feature set is going to be reasonably consistent across the marketplace - a result of the cutthroat competition among hardware vendors, and the maturity of the product base as a whole.

"If you look at the wide spectrum where sales occur, it's the value-add that becomes the real differentiator," says Rockwell's Bill Adkins. Ticking off some of those key value-adds: "Being able to plug and play in multi-vendor environment, dealing with multiple ACDs, having a single point of configuration, reporting and administration. This gives you a commonality of customer service across ACDs and centers," he says.

"Everyone has a degree of middleware. We each have different names for it, and then you have companies like Genesys that duplicate ACDs, but normalize across a multi-vendor environment. This brings in a high level of additional cost for the user. If you as a vendor can interface with competing products at the same cost, that becomes a value."

What this implies is that the ACD is no longer the central control point for a company's customer service strategy. It's more of a diffuse tool, albeit an important one, especially as alternate modes of interaction become dominant. If the ACD can continue to stretch to accommodate things like email and web interaction, then it retains some of its traditional role. If it ends up sitting alongside other "boxes" that route other interactions, then you see some of the role of the traditional call center extended outward to other parts of the organization.

That's where Murphy sees things going. "Clearly there's a broader set of corporate priorities. These things call center technology purchases are at the CIO level now - they're too fundamental to the business to leave to one department."

■ An ACD By Any Other Name

The ACD has evolved, like most telecom equipment, from a task-specific "dumb box" to a feature-rich intelligent device that can handle so many tasks that half of your other equipment becomes redundant. Originally, it functioned only as a call distributor, routing calls according to a database of handling instructions.

But after the landslide of technological invention that has characterized the last decade, the ACD has become a network-spanning piece of technology that manages all aspects of customer interaction.

By way of example (rather than endorsement), here's a taste of how ACDs have evolved in the form of one interesting product. Interactive Intelligence makes what could be called an uber-ACD, or a "call center in a box" — the Interaction Center Platform. The platform consists of two separate products, the Enterprise Interaction Center (EIC) and Customer Interaction Center (CIC).

EIC provides PBX, auto attendant, interactive voice response, speech recognition and unified messaging integrated with email (Microsoft Exchange, Lotus Notes, and Novell Groupwise). CIC adds in IVR, automatic call distribution, call recording, text chat, real-time supervision and screen pop integration with products from Pivotal, Siebel, Peoplesoft, and SAP.

Some of the things it can do include:

· Multimedia queuing for phone calls, emails, faxes, voice mails, text chats, Web callback requests and user-defined interaction types. Each interaction type can be assigned a "percentage utilization".

For example, with phone calls configured for 100% utilization, text chats 50% and emails 25%, an agent could be assigned one phone call at a time, or two text chats, or four e-mail messages. Combined interactions, such as one text chat and two e-mail messages (totaling 100% utilization) could also be assigned. Plus, one agent could be configured to handle only telephone calls, while another agent could be set up to handle phone calls, text chats, and e-mail messages.

· A supervisory alert lets contact center agents click a button to notify one or more supervisors when they encounter a problem situation. The supervisor can then listen in on the call, record it, or take it over.

· "One-number follow me" lets employees specify multiple forwarding numbers to be tried either sequentially or simultaneously when someone dials their extension. Users can also see who's calling and decide to accept or send the call to voice mail.

· A browser-based software phone through which users can control their calls, monitor queues and access company directories.

· A Palm-based software phone that lets mobile employees set up conference calls, receive alerts about calls arriving at their office extensions, and view the status of other employees. (This is weird, but somehow seems like the wave of the future, and is probably pretty easy to implement given the openness of the Palm OS.)

· Multi-site capabilities that let employees at one site see the real-time status (e.g. on the phone, at lunch, in a meeting) of employees at other locations.

The apps can be run on Intel/Dialogic voice processing boards, Aculab voice processing boards and Cisco AVVID voice-over-IP systems, to name a few. This open architecture is important because "interaction management" feature sets are going to be pretty similar across the industry, driven by some pretty fierce competition. What's going to set these devices apart is how reliable they are and how well they work with the other pieces of equipment you have in your back room.

Looking at the Interactive Intelligence suite leaves us with this question: Is there any hope for the good old standalone ACD? With the switch so far distributed out of the box, it's hard so see how a company could turn a profit just selling the box.

The key ingredient that drives the modern call center is the software — call direction, assessment and prioritization of calls and customers, reporting, and so on. What the modern switch has evolved into is nothing less than a software-based traffic cop, weighing the company's routing tables — not just of what agents are available, but also of the criteria to be assigned to customers, agents, problems, and interaction modes.

QUICK TIPS

1. There's no quick way to buy an ACD. There are, however, lots of examples of RFPs and sample proposals floating around on the Internet. Use as many resources as you can to try to distinguish features that seem innovative but have no real use in your center, from those that you'll really need. It might take time and be a torturous process. But with the cost of these systems, you'll be glad you didn't rush.

2. Consider the secondary market. The life cycle of a standalone ACD is shortening, from around six years to as little as three or four. As the switch's features move toward software, good quality basic hardware is more and more available from used dealers, at reasonable prices. That includes the critical peripherals, like phone sets. Even though it's likely that the switch of the very near future will almost always be spawned off a standards-based PC, there's still life in old switches, if you can buy them cheap. And there are plenty of people with the specialized knowledge to drive them floating around the industry looking for something to do.

Chapter Six

Outdialing Systems

As an outbound call center manager or supervisor, you get more than a little annoyed when your agents can't reach the people on their call lists. You know it's not your agent's fault. Much of their time is taken up trying to get through to a prospect to make a sale or collect a bill, and the longer it takes them to do their job, the more it costs.

Even if you have a small call center, a typical agent only reaches 25 to 35 people per 100 attempts, which could take hours. Enter predictive dialing: automation provides the same 100 calls in about 90 minutes, routing your agent only the ones that reach a human voice.

Today's dialers are much more sophisticated than they were fifteen years ago. Predictive dialing automates the entire outdialing process, with the computer choosing the person to be called and dialing the number. The call is only passed to the agent when a live human answers.

Predictive dialers screen out all the non-productive calls *before* they reach the agent: all the busy signals, no-answers, answering machines, network messages, and so on. The agent simply moves from one ready call to another, without stopping to dial, listen, or choose the next call.

True predictive dialing is merely one kind of automated dialing — there are others; but predictive is the most powerful and the most productivity-enhancing. True predictive dialing has complex mathematical algorithms that consider, in real-time, the number of available telephone lines, the number of available operators, the probability of not reaching the intended party, the time between calls required for maximum operator efficiency, the length of an average conversation, and the average length of time the operators need to enter the relevant data.

Some predictive dialing systems constantly adjust the dialing rate by monitoring changes in all these factors. The dialer is taking a sort of gamble: knowing that these processes are in motion, and knowing that there is a certain chance that a call placed will end in failure, it throws more calls into the network than there are agents available to handle them, if all the calls were to succeed.

Sometimes the prediction is wrong, and there are fewer failures than expected. In this case the called party will pick up the phone, say hello, and be hung up on when no agent is available. One of the intricacies of predictive dialer management is fine-tuning the aggressiveness of your dialer's algorithm.

Predictive dialing has been nothing short of revolutionary in the outbound call center. When agents dial calls manually, the typical talk time is close to 25 minutes per hour. Most of the rest of that time is non-productive: looking up the next number to dial it; dialing the phone; listening to the rings; dealing with the answering machine or the busy signal, etc. Predictive dialing takes all that away from the agent's desk and buries it inside the processor.

When working with a predictive dialer, it is possible to push agent performance into the range of 45 to 50 minutes per hour. I've heard of centers going as high as 54 minutes per hour. (You can't really go higher than that, taking into account post-call wrap up time.)

There is more to the technology than just the pacing algorithm. Predictive machines excel at detecting exactly what is on the other end of the phone, including the ability to differentiate a human voice from an answering machine. They typically decide that the call has reached a person within the first 1/50th of a second — the start of the word "hello."

Here are just a few of the important ways predictive dialing systems can help you.

- They completely automate outbound consumer calling. That includes the actual dialing, assigning agents and controlling the list you call from.

You can run multiple inbound and outbound campaigns, and you can specify names on a list not to call. It also schedules automatic callbacks for nonproductive calls. Dialers let you set the parameters for the dialing algorithms to meet the needs of a particular campaign, like the percent of overdials the system sends out.

With collections applications, for example, you may not care if the dialer has to hang up on a "customer" if there is no agent available. You'll trade the customer's good will for a higher volume of calls. But for a sales promotion, you'd want to keep those hang-ups to a minimum.

- You can manage your call center more effectively.

Standard features include real-time statistics about how each agent, group of agents or list is performing. Also, trunk pooling, which reduces operating costs by processing both inbound and outbound over the same trunks.

- They reduce agent burnout and turnover. Just imagine all the tedium they avoid: finding the phone number, typing it in, waiting for the phone to connect and the number to ring.

The dialer makes sure that the only calls an agent has to deal with are real calls, with a live customer on the other end. No busy signals, no endless ringing, no answering machines.

Cutting out that stalling doubles the time spent talking on the phone. Talk time, which is about 20 minutes an hour without a dialer, jumps to 40 to 50 minutes with one. Agents like their jobs better when they don't have to wait around for the phone to be answered.

- Reach more people in less time. You penetrate lists more deeply in a fraction of the time.

Predictive dialers adjust the balance of agents from one list to another, taking into account factors like list performance, time of day and the success of particular agents.

■ But Wait - Aren't They Complicated?

Yes, and no. The story of dialing in the last few years has been one of a technology that matured, and then was overrun by changes in technology outside the box.

By that I mean that the basic functions involved in predictive dialing (or any other dialing, for that matter) were long ago created and encoded into software. The rest of the cost of a predictive dialer was the cost of the high-powered dedicated box needed to make it happen, and to integrate it into the list system, and to the agent desktop.

It wasn't so long ago that predictive dialers were a simple purchase — you bought the one that gave you the most talk time per hour, or the one that had the best answering machine detect. What you looked for in a dialer was dialing features. That's changing immensely.

Like most other hardware technologies, predictive dialers are responding to changes in the nature of the call center. Nowadays you want more flexibility with your agents, inbound or outbound. You want to link your hardware systems together: switches and computers, dialers and voice systems.

More than anything else, you want to choose the software applications that make sense for your business, and get cost-effective hardware to run them. Decoupling the software apps from the hardware is the most impressive development to come along in years.

Predictive dialer vendors, like PBX and ACD vendors before them, have been forced to adapt to a changing world. People are less inclined to choose a standalone system they can't program and that can only be connected to a limited range of compatible peripherals.

Predictive dialing has always been a software application. It required a great deal of processing power, so vendors put their specialized software onto high-powered computers, most of them with a closed architecture. But the research and development was always geared to better dialing algorithms, more sophisticated call tracking features, and better database management — essentially software apps.

What started as a great idea for outbound telemarketing and collections — fire out more calls than necessary to maximize agent productivity — became the platform on which software companies continued to refine and develop new features for handling calls.

It was such a good idea that companies in other areas (telemarketing software, especially) began adding predictive dialing modules to their systems. The logic was good: if dialing features are mainly software, and powerful generic processors are available to run them, there's no reason not to create a whole new category of product — the PC-based dialer.

The traditional hardware/dialing vendors are now changing to match. Several of them have taken their core technologies, enhanced them, and are presenting them to call centers in a new light. They are creating systems for managing all aspects of the call flow. They let agents make calls in predictive mode, and receive incoming calls as well.

To facilitate that, dialer makers have incorporated a technology to blend agents; this allows a single station to handle either incoming or outgoing calls. And although it's not used widely yet, it's growing. The dialer is steadily losing its identity as a purely outbound object. It's got to act like, and interact with, inbound call routing systems. Because it's increasingly unlikely that a given center will be doing all of one kind of calling, or all of another. Recent information from Datamonitor suggested that the market for outbound dialers was actually expected to increase in the next few years.

Since few call centers are now dedicated to outbound traffic, integration with inbound is the highest priority for the vendors of high-end outbound dialers. Their strengths is clearly in the software that routes the calls, downloads the lists, tracks the results and coordinates the customer information on the back-end. If this sounds an awful lot like the new CIS software, you are right. If it sounds like computer telephony integration, you are also right.

The most successful predictive dialer companies right now — the ones making the most interesting and useful technology — are the ones that have rethought the logic of the outbound call center and recast their dialer as an indispensable component of the inbound and outbound center. For all of them, the selling point is not the power of the dialing engine, but the value-added capabilities of the companion software.

■ So Predictive Dialers Head For Software...

I found one company that put together a custom installation, using their own programmers, with a software-based predictive dialing system and Dialogic boards. It may not be the best system for everybody, but it is definitely possible to get predictive dialing for less than you expect.

Dialing *is* by definition software. For years, the predictive dialing vendors (rightly) competed with one another on features — answering machine detect, speed of answer, and fundamental algorithm — that were software. The boxes were of secondary importance. They were proprietary because you needed lots of processing horsepower to drive those software applications.

Nowadays you want to have more flexibility with your agents, inbound or outbound. You want to link your hardware systems together: switches and computers, dialers and voice systems.

The logic behind it is overwhelming: if dialing features are mainly software, and powerful generic processors are available to run them, there's no reason why they can't be part of an overall inbound and outbound call routing system on a client/server platform.

So what should you be thinking about when buying your predictive dialer? Integration — with every other piece of hardware and software in your call center. Mostly software.

What's happening in the call center now is the marriage of voice and data. Call centers are using open dialing platforms to take advantage of other niche technologies in the call center. It's a powerful means of taking the benefits of predictive dialers even further.

Why is integration important to call centers? Primarily because of the increased control call center managers have over their technology. Essential call center equipment like ACDs, PBXs and predictive dialers now work in concert, allowing for greater efficiency and productivity.

Companies are driven to make better use of their resources. There are so many technology directions that companies can easily fritter away resources and not really improve the service they provide or the bottom line that they protect.

Smaller centers have basically three options:

1. Buy a turnkey system (which may be proprietary) and build a telecom and computer system around it. This is good for companies that want to dump older equipment.

2. Go for an integrated solution, combining the power of PCs and LANs with software or hardware dialing processors and a phone system.

   Using off-the-shelf parts, you can put together inexpensive solutions. You can grow into it slowly, without sacrificing the dialing features you need: swift answer detection and screen transfer.

3. Or lay a dialing solution on top of the existing telecom and data infrastructure. Your best option here: talk to the vendors who make your existing equipment and software. Chances are you might find a dialer maker among the vendors of your ACD, VRU or call management software.

Software-driven predictive dialers integrate into the call center environment because they're based on multipurpose minicomputers that let users run other software and because they employ industry standard computer-telephone devices to perform predictive dialing.

Besides predictive dialing, many dialing systems let call center agents perform preview dialing where agents call up data and review it before the call is placed. Preview dialing mostly benefits small call centers making business to business calls.

As for the future of predictive dialers, most agree about the importance of integration. For some call centers, integration means less dependency on mainframes, while others see it as a way to tie dialers into a national database of people who don't want calls. Even more adventurous is the theory that full function predictive dialing will be possible from an agent's home phone.

Regardless of what happens in the future, one thing is true: forward thinking has turned a once limited piece of hardware into a versatile and vital piece of technology. As long as that persists, its value in today's (and tomorrow's) call center remains undiminished.

▩ What to Look For

· Databases. The dialer must connect to your host system to access your call lists. It's got to switch between lists and campaigns and it's got to reschedule callbacks for busies and no-answers.

· Continuing improvement in software and applications. With a clear focus from the vendor on integrating the dialer with inbound switching and with the total customer management software infrastructure. It's not going to do you much good if a sales person works from one isolated island of information, and the customer service department works from another.

· Call blending. You don't have to divide agents into inbound and outbound pools anymore. Once, it was necessary. But why should you have outbound agents rushing through a call list while inbound agents sit idle?

What happens when an automated outbound center starts to receive callbacks from the people not reached on the first call? (That happens a lot in collection environments.)

If the ACD and the dialer don't communicate well, the calls could go to an inbound-only group. But you may want the same agents to handle inbound and outbound. Or a particular account may belong to an agent or a group.

Intelligent features are available for blending inbound and outbound calls on the dialer. You adjust for peaks and valleys by dynamically switching agents from one group to another. It reduces staffing inefficiency and maximizes agents' talk times more than ever before.

There are two kinds of call blending: reactive blending, where agents are switched around when the system detects an overflow; and predictive blending, in which you predefine when the pools are switched.

· Computer/telephony integration. Dialers should be linked to the host telephone switch, as well as fax and voice mail systems. Monitor ACD activity through server-based connections to the switch. These reports help control workflow in a call blending situation. All this information is vital — how many calls come in, how long they are on hold and when the call center is busiest.

· Business-to-business impact. Predictive dialers are great when you have large lists of customers or potential customers. But they don't work well when you have to call businesses.

That's because companies almost always answer their phones, even if the specific person you want to reach is not there. You lose the efficiency gained by screening out no-answers and busies. The same problem applies to people whose job it is to screen calls.

If you need to do this kind of calling in conjunction with consumer telemarketing, consider a dialer that lets you shunt several agents into a separate campaign, then run that campaign with the predictive features turned off. Lots of dialers let you do this.

If you do this kind of calling exclusively, try a sales management software program that controls your lists, feeds the sales person the phone numbers, but lets them keep control of the call from dial to hang-up. This is called preview dialing and it's common to most contact management or sales automation software programs.

· Do a 30-day test of the dialer with each vendor you're considering. With a 30-day test, managers will experience not only what the system can do but what impact it will have on your people.

· Make sure the system supports the number of users you need now and in the future. Take a close look at the stability of the company — those that made just dialers in the '90s have been dropping like flies.

· Beware of slow transfers. If the prospect has to say hello three times before the call reaches the agent, you might lose the sale before you've made the pitch. Your system should transfer the call fast enough for the agent to hear part of the first "hello."

· Beware of call abandonment. Some lower-end predictive dialers hang up on 20% or more of prospects. The last thing you want is to generate complaints: for people with Caller ID to be calling your center, or calling the phone company, or the state regulators. List owners who receive complaints may stop renting you their lists. Try to buy a system capable of a zero abandon rate. This is a necessary precaution in case future regulations require it.

· Get a system that's capable of inbound/outbound call blending. That's the ability of every agent to handle both incoming and outgoing calls from the same station. This leads to higher list penetration and more sales. The outbound agent can leave a message explaining what the call is about. Calls that come back in based on that are more likely to end in a sale. Likewise, inbound-based agents can handle outbound calls during slow periods.

· More leads can mean a lower sales conversion. Perversely, when the supply of fresh leads is unlimited, agents sometimes don't try as hard for the sale as they do when the supply is limited. Especially with very little wait time between calls.

· Get call tracking and custom list loading. Your system should keep a record of call attempts to be certain that each attempt is at a different time during the day and on different days. Some weaker systems have no call tracking abilities at all, and every time you load the dialers the same people are always called first.

· Keep transition times short. Predictive dialing should reduce the after-call work time (that time between when your agent hangs up the phone and when they are back on the phone) at a given low abandonment rate, say, 2%.

Only a few predictive dialers can achieve list penetration level of 50% with daytime calling, while maintaining an average wait time between calls of nine to 20 seconds at a 2% abandonment rate.

· Get full branch scripting. The best scripting systems allow data and calculations to be performed in the script and the screen routing is based on the answers given.

Also when an agent hits a key to change screens, it must be instantaneous, no matter how many people are on the system. A multiple-second wait time is unacceptable. The system should also provide function keys for access to questions most likely to come up at each specific place in the script.

· Ensure campaign flexibility. There should be no limit to the number of campaigns or number of projects that can be called. Every operator should be able to call a different campaign at the same time.

Make sure it's expandable. What if your company becomes a big success? Plan for future growth. (I always have to say this, but it's true.)

▓ Don't Forget About The Rules!

These days, all anyone asks about is Do-Not-Call: the set of regulations that set up a nationwide registry of numbers that are forbidden for outbound telemarketing in the US. This, despite the fact that the call center industry in North America is predominantly an inbound one. Why is that?

I suspect it's because the DNC regulations scratch the itch that irritates consumers the most, the annoyance of home-bound telemarketing. And that the media, which loves to play mind games with an annoyed public, has latched onto the DNC movement as a way to talk about all the ways companies harass consumers.

But the truth about the DNC is much more subtle, cynical, and pathetic.

Truth alert: I put all my phone numbers into the do-not-call database. I don't want to be called at home by people offering to sell me credit cards or long distance service. Sorry, I already have enough of both. It was an empowering feeling; for just a fraction of a second, I had the wonderful feeling that I would be taking back my personal space from those who call me at dinner.

Having done that, I expect that the net practical result of the DNC on individual telemarketing will be nearly unnoticeable.

The DNC rules include adequate loopholes so that anyone who really wants or needs to call you will be able to justify doing so.

What it really does is create a compliance standard that allows the telemarketing industry to assert (with justification) that they are following honest business practices.

Those that choose to flaunt the rules will be able to do so without much fear of reprisal.

And in the end, most consumers will continue to receive calls from both those that comply, and those that don't, and typically they won't be able to tell the difference.

Leading to the conclusion that the rules themselves - cynically crafted to exempt the very politicians who drafted them - are a smokescreen and a won't change business behavior or make a dent in consumer irritation.

I could go on, but why? Some in the industry want to fight the rules, which is silly, because the rules don't prevent telemarketers from practicing their business in almost exactly the way they have always done. And those companies that give everyone a bad reputation, the boiler room outfits and scammers, they will continue to operate beyond the margins of good taste and good behavior regardless of the regulatory environment.

And in the end, the consumer will be frustrated and disappointed because he/she will continue to receive calls at annoying times for products they will almost always not want - and they won't understand the difference between those who comply and those who do not. Nor will they care. They'll just look at the entire universe of people who call them as "telemarketers" and vent their anger at the whole industry, inbound as well as outbound, compliant vendors as well as non-compliant.

But the bottom line remains: DNC is the law, and you must be in compliance. Your dialer vendor can/should/must give you information about the ways in which their equipment can be used correctly and effectively.

■ New Tricks For An Old Tech

A new approach to answer-detection hopes to take some of the "nuisance" sting out of outbound telemarketing calls.

Outbound dialing may seem like old hat, but despite the power and maturity of today's dialers, some problems remain. People still hang up when they hear dead air, or see an unknown number on their Caller ID display.

One company, Castel, says that as many as 60-70% of customers hang up or ignore calls coming from outbound centers because of that kind of problem. Castel has been working with several call center-based companies including, Omaha Steaks, Haverty's Furniture, and Verizon, to develop a new call center technology that reduces telemarketing nuisance factors from outbound calls.

They are just out with a tool they call DirectQuest. It's different from traditional predictive dialing systems, they argue, because it incorporates technology that communicates directly with the public switched telephone network at the command and control layer. And they content that this approach to dialing guarantees answer detection 100% of the time.

By exceeding FTC, FCC and state regulations, call centers using DirectQuest are also free from outbound call fines as they try to keep up with compliance needs.

"Today's outbound call center industry is in crisis," said Geoff Burr, Castel's President and CEO. "Consumer backlash and regulatory restrictions against nuisance calls are striking at the heart of the call center industry's ability to stay in business. The cause of this crisis is a predictive dialing technology that takes too long to recognize that the phone has been answered, because it's looking for noise energy. This is why Castel's DirectQuest solution is so timely; it reads network signals instead of measuring voice energy."

"Digital Signal Processing (DSP) technology, which is the foundation of every other major predictive dialer on the market today, determines the result of each outbound call attempt by analyzing, or listening to, patterns of noise energy and silence," he says.

When a signal pattern meets the criteria set by the manufacturer for an answered call - such as the presence of noise energy followed by two seconds of silence - the call attempt is considered a connect and is routed to an agent.

DirectQuest differs in that it looks for the off-hook message that is sent back from the answered phone to the originating Central Office Switch. This message is the same message telephone companies use to initiate billing. DirectQuest responds by routing the connected call and its appropriate data screen to an available agent. The amount of time it takes to receive this connect message and route the call to an agent is in the tenths of a second range. By the time the called party says "Hello" an agent is already there to return the greeting.

DirectQuest is a modular software application that works in conjunction with Castel's CTI interface and a the company's digital softswitch. The server-based application is capable of centrally managing the calling operations of a network of distributed softswitch dialers.

QUICK TIPS

1. **CAVEAT EMPTOR.** When shopping for small-scale predictive dialing systems, beware of the loose use of the word "predictive." Predictive dialing, by definition, is not available for one agent. It requires groups on which to base its assessment of how many calls to place into the network. A single agent (or even a small group) is not enough on which to base an accurate judgement.

   What some vendors offer as predictive dialing is really anticipatory dialing. What's the difference? It uses the statistics generated by a single agent to "anticipate" when that agent will be ready for the next call. It dials the call while the agent is still talking. If that call takes too long, then you have an abandoned call. It only takes one abnormally long call to throw off the stats. What you don't have with anticipatory dialing is the cushion of averaging talk times across groups of agents.

   Another term you might run across is power dialing. This is (roughly) a bulk dialer that shares some characteristics of a predictive dialer (good for small groups) but that does not have an algorithm for calculating ahead of the agents and predicting when one will be ready. It will, however, screen out busies and no-answers. If a vendor makes a dialer that doesn't have a predictive algorithm, but they want to position it in the marketplace as more advanced than a mere preview dialer, they may tag it as a 'power' dialer, confusing all and enlightening none. As the term with the loosest definition, it's rapidly losing all meaning.

2. Does your dialer let you cross-reference lists with other databases based on demographics, geocoding or other characteristics? It should. And you should be able to change lists on the fly and have these changes instantly affect the current project.

3. Make sure that you maintain a Do-Not-Call list, as required by Federal legislation and some states, and as recommended by most major telemarketing associations. Yes, this is for all intents and purposes a fool's errand — there is just about no way to make this work in practice. But it's better to be able to say you tried and missed a few people than to get caught not doing it at all.

◼ Telemarketing & Scripting Software

Telemarketing software is a major investment, but the amount of money saved through increased employee productivity allows the system to pay for itself — usually within the first six months of use.

No matter the size of your call center or the level of automation, the right telemarketing software can add productivity to your operations. Customers get better, more personalized and faster service when your reps use telemarketing software. Your agents have everything at their fingertips. They have time to make more calls. It makes follow-up much easier.

I'm talking about software that keeps contact lists, fulfillment information and buying history. It runs scripts and campaigns, most especially in outbound sales programs. With most packages, all information can be retrieved and updated by any of your reps. This means customers get the same service each time they call, regardless of what agent they talk to.

Most of these systems are now concentrating on taking the data out of the call center — allowing managers to track customer information enterprise-wide, and in fact there are some who believe that this niche is in danger of being subsumed into the newer and wildly popular *customer relationship management* category. There is some truth to that point of view.

However, there will always be a place, especially in the small center, and those that are dedicated to outbound calling, for software that simply manages the list and the dialing. In this case, you are less concerned with front-end/back-end integration than you are with the outbound front-end and crunching as far through a list as possible.

Here are some of the things you should think about when considering what kind to buy.

Scripts. One of the best ways to break in a new sales agent is with a script. You can create scripts for different telemarketing activities like sales, follow-up or customer service.

Then, agents who move from one group to another (or cover for another group on short notice) have a way to handle calls without sinking. It's also the most critical way to enforce a consistency of approach — making sure that all components of a telesales program are brought across to the customer every time. Otherwise, there is no way to accurately measure the productivity of either the program, or the sales reps.

Software makes creating scripts a matter of flow-charting the answers to various questions. Branching logic is standard on most major software packages. From the agent's point of view, he or she reads the appropriate question off the screen, and selects from a menu of choices depending on how the customer responds.

Increasingly, "telemarketing" is just one function of much larger application suites aimed at call centers of various sizes and types. This makes sense, when you think about it. From a software developer's point of view, "suiting" allows them to expand the potential market for their products; if one particular application falls out of favor (as telemarketing software appears now to be doing), they can de-emphasize it when selling the suite. Those centers that still need it will find it bundled into larger, more expensive packages.

List management. Telemarketing software's other defining function is the ability to create a calling list from information in your database. You can use telemarketing software to clean your existing calling lists. Some programs are able to merge/purge out useless names. Or you can run the lists against larger databases to flesh out missing information.

At the very least, a good telemarketing program ought to perform day-to-day maintenance of the company's in-house lists: customers, leads, sales, fulfillment. This goes hand in hand with a program's database features.

Sales pipeline analysis. One of telemarketing software's most important features is the ability to give management instant feedback. How many calls turn into sales? What is the return on investment in the call center?

Managers want to see how these programs are working to produce sales, so they can correct scripts or change the segmentation of lists. The software you choose should let you analyze how much time and money each account costs at each stage of the selling process. That helps the call center focus on revenue generated from each sale, rather than just on the number of calls made or contacts reached.

Handle inbound calls, outbound calls or both. You may want to run simultaneous campaigns. Or feed the results of inbound-generated leads into an outbound calling cycle. If your software doesn't have this capability, it doesn't have much. At the very least, you should be able to switch reps between handling inbound calls and making outbound calls, and you should be able to see — instantly — all queues, surveys and scripts tied to campaigns.

Database compatibility and maneuverability. Even if you are just beginning to automate, finding software that works with a number of database formats is vital for sharing information with other departments, other companies, even for importing list information.

And while you hope it never happens, if your vendor goes out of business, or if you decide to switch systems, you don't want to lose your investment in your database too.

Transaction processing and order fulfillment. Your outbound campaign is going to be most productive if the scripting system that convinces the customer to place an order segues directly into a system for placing and fulfilling that order.

If you've gotten the impression so far in this discussion that this category is a little stale, you're right. Telemarketing software's heyday has been here and gone, especially now that outbound almost always goes hand-in-hand with inbound telesales, direct response, and customer service.

There are few if any companies that make software that is only for outbound telemarketing, and in many ways the category just below that — contact management and sales force automation — has pushed its way up. Programs like Goldmine, once just a simple tool for helping a single salesperson get through the day, have evolved into full-blown network-based, multi-user sales tools.

Companies that specialized in the list management and scripting areas have by and large reached upward to expand into CRM, leaving this space largely empty of products that do this and this only.

If you want something that will see you through several generations of growth, and that you can do very quickly, you may want something more along the lines of the suites available from Cincom, Peoplesoft/Vantive, and Pegasystems. There are lots of companies in this arena, depending on the vertical industry you occupy, your price sensitivity, your sensitivity to having consultants on premises to set you up, and other factors.

Also note that if you have (or are considering) a predictive dialer, the choice of software may be a complete non-issue — it may already be part of the package. You can get some good software with the dialing engine. The dialing companies don't just make high-powered hardware. Many dialers are now PC-based and come with snazzy telemarketing software. Look at what's coming out of Concerto, Aspect and even Oracle (some time ago it bought a company called Versatility, which made one of the original PC-based telemarketing and dialing systems).

Concerto's system, for example, offers features like branch scripting, the ability to make changes to campaigns in progress, a report writer and *Enhanced Recall Queuing*. This feature lets reps remain available for calls on completed campaigns while working on new campaigns.

Computer Telephony

Any company's main focus should be its customers: fielding their calls, delivering service, getting orders out the door, making sales. The easier it is for a customer to get in touch with you, the better the relationship will be. Companies that do the best job of opening the door to customers, those that make it as easy as possible for customers to find out what they need to know, are the ones that have the best track records in the long term. Small and medium-sized companies that have adopted customer-focused attitudes have, over time, become giants of their industries.

Over the past few years there has been much discussion of the pros and cons of a set of technologies called CTI, or computer/telephone integration (or just computer telephony — they all mean the same thing). Computer telephony was designed specifically to enable better contact between companies and their customers.

It is a loose but complicated amalgamation of interlocking technologies. It isn't any one thing, not any one piece of hardware or software. It's a way of combining the two streams of information — voice and data — through open, standards-based systems. It has uses in all areas of modern business, but its most dramatic possibility is in the call center. If implemented well, it can improve the way a company interacts with its customers, which of course, is the whole point behind the call center.

Computer telephony is a way of reaching beyond the traditional limitations of either of the component technologies (phones and computers) and bringing them together in a way that improves them both, by bringing more information to the person on the phone, and making the data behind the scenes much more flexible.

Think about it. The ability to integrate your computer and telecom system could bring the customer's phone call along with his datafile right to the agent's desktop, as the call comes in. This translates to massive savings in 800 line charges and agent labor.

In practice, implementing computer telephony has been a dicey proposition. Until very recently, it was largely custom, with each venturesome company taking the plunge using a systems integrator to cobble together all the necessary links, proprietary interfaces and special connections to applications. The benefits are easy to see, but sometimes difficult to achieve. Most call center CTI experiences begin with good intentions. Somehow, they don't all end up that way. Imagine this scenario:

Your company is facing stiff competition and is growing rapidly, resulting in a certain amount of customer tension — people have a hard time getting you on the phone when things go wrong. It takes too long for sales reps to respond to good leads. Emails come in from customers and go...you're not exactly sure where. Same thing for fax traffic. You don't even have time to think about the traffic coming in from your website (and whether your site is connected to your call center).

You hear that there are technologies out there that promise relief. They promise to tear down the walls between you and your customers by bringing voice and data together. You swallow the bait. You hire a consultant and they present you with a plan. Screen pop, says the consultant. When a call comes in, shouldn't the agent have all the information? Sounds good, you say. Single point of contact, he says. So when a customer calls, whoever handles it has all the relevant info. Makes sense, you say. Links between the switch and the host. Connections everywhere.

Of course it makes sense. And before you know it, you are in the middle of an implementation. The months drag on. The consultant puts a dollar figure on the technology, but once he's gone from the scene you realize that his number didn't include things like training, or coordinating what happens in the center with what goes on in other departments. Of course, the technology works, but do the people know how to work the technology?

A year later, you are staring at the prospect of starting all over, with a different set of technological priorities, a different consultant, but the same basic feeling in your gut that yes, you do need to get closer to your customer. You just need a better way to get from Point A to Point B — one that has clearly defined cost and benefit signposts along the way.

Installing CTI systems used to be an incredibly custom job that involved detailed on-site "fixing" to make sure that everything worked together. Luckily, things have changed a lot.

Computer telephony is simply defined as "adding computer intelligence to the phone call." When you think of it that way, everything from simple screen pop to predictive dialing becomes, at one level or another, a computer telephony application. Depending on your call center's level of sophistication, and the capabilities of the underlying telecom infrastructure, you may already be using core computer telephony technologies.

■ Where Did CTI Come From?

Consider, for just a moment, the historical anomaly of telecom. Despite the conventional wisdom about the rise of the microprocessor and the computer revolution, the fact is that the national telephone network built incrementally by AT&T over the course of decades was a feat of computational and networking engineering unmatched in the 20th century.

And yet, when microprocessors begat PCs and PCs begat client/server networks, the companies that made the phone switches for average businesses remained curiously unmoved. Notwithstanding the fact that the business phone system is one complex piece of computational hardware, there was a great deal of resistance to making the phone act more like a computer.

The computer, though, was easy enough to make act like a phone. The computer industry was better at implementing the things that make disparate technologies talk to each other — important things like vendor-independent standards.

Computer telephony has its origins in the fact that if you wanted to add on to a typical office PBX, you had to buy the add-on from the original vendor, or from a third-party company that wrote to the proprietary spec promulgated by that PBX vendor.

Good applications were hard to find because for a software company writing these add-ons, the cost of developing for multiple switch vendors was prohibitively high. If you wanted a reporting system to complement your switch, you had only a few options: buy from the vendor or the vendor's approved partner, or build it yourself. None of the options were particularly attractive.

Computer telephony was an attempt by the more perceptive members of both the PBX and computer industries to come to grips with the notion that they were more alike than different. Switches were really high-performance communications servers, after all. If the specs could be opened up, if standards could be developed, both sides would benefit from the flood of applications that would be developed.

Today's switches come with computer telephony hooks built in, and a suite of applications from the vendor and its partners that take advantage of the connections.

▧ CT's Changing Face

What began as a tentative effort by PBX vendors to open their switches has turned into a more solutions-based set of technologies.

Part of that is due to the widespread adoption of technical standards for interoperability between vendors and industries. At the basic level, there are standards for the operation of component hardware at the board level. There are also specs put out by individual vendors that enable applications to work correctly on particular board sets — SCSA and MVIP, for example, by companies like Dialogic and Natural Microsystems.

There are also standards, created largely by the computer software industry, for the creation of applications that work with operating systems. The key standards, TAPI and TSAPI, were offered up by Microsoft and Novell, respectively, as a way to push the switch vendors into compatibility so that developers could use those companies' OS platforms as the basis for CTI applications.

Some of these apps focused on call control (the movement and tracking of calls around a phone network). Many others were apps that took advantage of the growing LAN/phone system connections to bring data to the desktop at the same time as the phone call arrived. Wherever voice and data networks come together, you need standards to assure that the integration goes smoothly (or works at all).

The Internet, of course, forced things to become even more standards-based. Building apps that combined call control and data manipulation became a lot easier with the adoption of Java, TCP/IP and ODBC as standards for data communication.

While most apps assume that there will be a dedicated piece of hardware for the pure telephony switching, it's becoming clear that nearly all the add-on functions of value — the call center specific applications — can reside quite nicely in a "telephony server" hooked into the phone switch. More and more, that telephony server is a Windows NT box.

In addition to raw interoperability standards, there are more different kinds of links to consider. In any data/voice application, there are a lot of possible combinations. When I say "voice," for example, I could be talking about a lot of different kinds of "calls" — traditional phone calls, for one, but also recorded calls in the form of messages, fax traffic, even the digits that callers enter when they pass through a voice response system.

As for data, this starts with the host information that resides in the back office databases. But it also includes the subset of host data that moves to the desktop (and back). And MIS data that passes through the corporate LAN, through intranets, over the Internet (including company Web traffic), and emails.

It used to be easy to isolate the two streams — to break it apart and clearly identify what was voice and what was data. Today, though, a corporate CTI system might also be dealing with strange combinations that include elements of both sides: things like voice-over-the-Internet, fax-over-the-Internet, browser-based transaction processing, "call me" buttons that appear on Web pages. Even speech recognition — all these combos are the result of standards that make it easier to push data and voice across each other's pathways, and that make it increasingly irrelevant in what form a piece of information comes in. What's more important is how that information is acted upon, and who has access to it.

CTI is now so broad that it is best defined as any technology that combines some form of real-time, person-to-company communication with a background of data that adds-value to that communication.

When I first wrote about this, in the first edition of this handbook, computer telephony was a developer's technology. It was still being thought out, and built into the core telecom. It needed to be embedded in the switch, or the switch needed to be open to one or more of the "glue" products that sit in between the switch and the applications you wanted to run ("middleware"). Now, that integration is showing up more behind the scenes, as products (particularly applications) for call centers assume a greater degree of CTI readiness on the part of the switch and the center infrastructure as a whole.

The phenomenon of switch-to-host integration represents a total transformation of what you can do with a center. Thanks to this category of product, the most sophisticated call center features are no longer only available to the biggest, highest-volume centers. Small companies can now avail themselves of once prohibitively expensive technology, taking advantage of ANI, DNIS and other network-provided services to do a lot more with each call. This places them on a more level playing field with their most mammoth competitors.

Some of the most obvious benefits that computer telephony offers the call center:

· Have shorter calls. Cut hold time dramatically. Speed information to the agent's desktop, then to the caller. Reduce your telecom usage costs (the second biggest expense in a call center).

· Have happier customers. Simply put, your reps solve more of their problems the first time out. And faster.

· Make more sales. You have more information about the caller. And, more important, you can bring the information that's hiding in the corporate database to bear on that particular call exactly when it's needed. You know what they like to buy, and what problems they've had in the past. You can appeal to them on their terms. They don't get passed from agent to agent. And you can cross-sell or up-sell them while building their loyalty.

· Make better use of staff. Gain efficiencies through blending, and other ways of creating dynamic, responsive group configurations. Slow period for calls coming in? Move some of those reps to the outbound side. A dialer seamlessly starts sending them calls, a script pops on the screen, a whisper prompt in their ear tells them the name of the person they're talking to. Presto — no more down time. Now, it's not quite as simple as that, but you get the point; computer telephony puts more information — meaningful information, not just raw data — into the hands of the people who can use it most.

· Improve customer service. Put more information into the hands of the customers, with or without agent intervention. Customers can often serve themselves. This costs less and frees up company resources for more complex tasks.

· Connect with the Internet, or with company intranets, with all sorts of multimedia sales and service tools. All of the traditionally expensive tasks, like order processing, literature fulfillment, interactive faxing, are made easier through computer telephony.

■ What it's Used For

Voice response systems are front-ends to the phone system that deliver recorded information when someone calls. Interactive voice response is two-way; it responds with information when a caller enters digits on the touch tone phone. And when that information comes from a host database, that's CTI in action.

Customers can call at any hour of the day or night looking for account balances or order status information. The IVR engine queries a database in the background and reads the information to the caller. In this way it can be made dynamic — instead of just reading off a set of canned, pre-recorded announcements ("we're closed right now, please call again tomorrow" or "to leave a message for our sales department, press one") it pulls real time data out of corporate databases.

When this is translated to the Web, the kind of information you can make available to customers is expanded dramatically. Anything visual, from catalogs to product schematics, can be dropped onto a customer (or agent) desktop. Customers can help themselves when problems arise. They can learn about your products before they buy. And when it comes time to talk to an agent, they are better prepared; so the call is shorter, more effective, and more profitable. The "shopper" does his shopping without consuming your most valuable resources. But the buyer gets your full attention.

CTI is an information delivery tool. It won't make a seller out of someone with no sales ability, but it will give someone who deals with customers the knowledge they need to address the needs of the customer.

For starters, a customer record is brought to the agent's desktop at the same time as the customer's voice arrives on the phone. The caller can be identified in many ways, including the information that travels with the toll free call, or through digits entered by the caller himself. When the agent has the customer information in front of him, the call doesn't last as long. The customer doesn't have to repeat himself every time the call is transferred. And the agent sees the entire history of the relationship with that customer. If the customer has a history of problems, the rep will know about it. And if the customer has a million-dollar lifetime value to the company, the rep will know that too.

Better still, if that caller is really a million dollar value, the CTI system will know it before the rep will, and can be set up to send the call directly to someone equipped with the experience to handle priority customers.

Another way CTI helps is with quality control. In call centers, calls are often monitored — recorded and archived so that the agent and his or her supervisor can listen to them later and assess performance.

That analysis process is made much more productive when it's augmented by the data that passes through the agent's screen during the call. A complete record of every transaction can be kept indefinitely, including every screen viewed by the agent. This "screen scrape" is an audit trail and a training tool.

All these things help a company cut operating costs by reducing (or stabilizing) support staff headcount. A company can be more productive with the same staff by handling more calls (or customers, or transactions, depending on which metric is most important to them).

And most important, it lets a smaller company look like a big one — without sacrificing the personal touch. Customers don't care how big a company is, they care about what kind of response they get when they call or contact that company. If they get good service from Federal Express or L.L. Bean, they're going to learn to expect it from every other business they deal with, no matter how large. CTI applications allow companies to appear more fully outfitted than they really are. This could mean putting systems in place to answer calls during off-hours when no agents are available, or having a website take orders at all hours. In any of these cases, the underlying technology that links the computer networks and the telecom systems enables the small company to decide for itself how to manage its customer relationships.

A call center that uses computer telephony knows who its customers are and why they are calling. It knows what they like, what they dislike, and how much they are worth to the company. On the other hand, without computer telephony every customer interaction is like a blind date — full of expectation, and possibly, frustration.

CTI lets a company respond faster to changing market conditions. But it must be implemented correctly: with clear and ongoing support from upper management and a clear-eyed view of the company's goals for the technology.

For a company to put computer telephony into place requires that they determine, from end to end, exactly what they want a customer interaction to be like. Every contingency must be accounted for: phone calls; emails; fax requests; even Web hits. Far too many companies have had disappointing results because they didn't put in the computer telephony they *needed*; they put in the technology they imagined. The right CTI is the mix of applications and core technologies that add value to the company's existing operations, and allow it to do more: voice mail, unified messaging, advanced call routing, fax redirection, Internet telephony, call center apps, customer service software, sales force automation — whatever combination is most useful in their particular circumstances. The key is to figure out which pieces are right for which circumstances.

There are many ways to make it work. There are also many places to go for expert advice. Component vendors start the process, and often point the way to application partners whose product works with the core pieces. Telcos and other large service providers can also provide an umbrella under which integration between all the pieces are certified to work correctly. There are systems integrators, who specialize in matching the various pieces to the custom needs of a particular company, or vertical industry.

Whichever direction, the growing company needs upper management buy-in, direction on the goals of the project, and a clear-eyed view of the relationship between the company and its customers.

▓ The Call Center Jigsaw Puzzle

Putting the pieces of a CTI system together involves an amazing degree of coordination between products and vendors at several levels.

The bottom layer consists of the fundamental hardware and conjunctive elements: the boards that process the voice and data channels; the servers and networks, often ruggedized to reflect the mission criticality of what they are used for; and the standards and open APIs that link different vendors' equipment together. The most common boards used in CTI systems are from manufacturers like Dialogic, Natural Microsystems, Brooktrout and several other specialty companies, depending on the application.

Parallel to that sits the dual networking infrastructures: the phone switches and the data networks. The phone switches are usually PBXs or dedicated high-volume call routing switches called "automatic call distributors," or ACDs. Phone service is also a core component. Not just because it's an obvious necessity, but because increasingly, the carrier networks are being upgraded to deliver advanced call processing services through the network. Sometimes this works directly to the advantage of the smaller business — if messaging or call routing applications can be run from the network, you need to invest less in premise-based equipment. You can implement "high touch" services like call centers without spending so much on high tech infrastructure.

The data networking infrastructure, like the phone system, is probably already in place: LANs, intranets, external Internet connections and websites, desktop browsers and firewalls.

Between these two networking areas lies the middleware layer. The products in this category are what most people think of when they say CTI — the very specialized applications that draw data out of host systems and coordinate it with incoming telephony information, then format it for both sides. Originally, many of these products focused on coordinating between a single vendor's switch and a single host format. As a rule of thumb, the older and more widespread the databases, the more important (and more complex, and customized) the middleware has to be. This has accounted for a lot of the tension surrounding the installation of CTI. For companies with decades-old legacy systems and extremely customized databases, installing CTI meant that to achieve any of the benefits, you had to go through a trying period of intimate customization between the switch and the database.

Increasingly, middleware connectivity is being sold as part of the switch, and middleware companies themselves are being acquired by larger companies above and below them on the CTI component chain.

The next level of product in the CTI hierarchy is the application layer. This is the software that actually does the things that make people more productive, things like messaging or speech recognition, automating sales forces or taking orders through the Web. It's a good idea, when pondering a transition to CTI, to start here, with a concrete idea of what you want the system to accomplish. It's akin to buying a PC based on what kind of application you want to run. You pick out the spreadsheet and word processor that has the features you need, then buy a PC that makes those features work best. CTI is no different. The best approach is to identify the applications that suit your business and then build up and down to integrate those apps with the infrastructure you already have.

Along with these layers of technology come the consulting services and systems integration know-how that ties it all together. For the most part, CTI is not an off-the-shelf accomplishment. It does require intimate connections between different technical realms — which are usually managed by different people, with different sets of priorities.

■ How to Make it Work

There's no way for me to tell you everything that can go wrong (or right) with a CTI install. There are so many things that can't be anticipated by outsiders, which is another key reason why you want to have an internally directed plan, rather than handing everything over to a consultant or a systems integrator.

Many companies need help defining the scope of what CTI should do in a business context (not just from a technical point of view). That help can come in the form of a consultant or systems integrator, who typically work with a company to coordinate the entire plan of an installation, help select the products from the various layers, and if necessary create any custom linkages or applications to suit the situation.

Or, that help can come from one of the vendors themselves. This is more common than it used to be, as vendors have worked hard to orient themselves in a way that makes more sense to the end-user: end-to-end coverage of the entire CTI process, from the component layer through the applications and service. They often set up umbrella offerings through application partners, so that a customer can choose from a variety of apps that are all pre-certified by the hardware vendor to work together.

So use this short list as a tickler, to spark some thoughts about the kinds of things that applies to your particular business circumstances. These aren't the only things to watch for and pay attention to; they are just the ones I hear about most frequently.

· Single out the high-volume areas of your call center operations. For a telemarketer selling stereo equipment, the customer service and new orders division may field many calls, while the help desk may get a relatively low load. For a PC vendor, however, the help desk may be just as inundated, or more so, in the face of decreased sales.

Before implementing any computer telephony technology, you should define the internal environment. The areas with high volume are going to have the highest payback when you implement open applications. Possibly, only one or two of your applications would really benefit from computer-assisted telephony.

The telecom manager should also assess the nature of each department, to see if the switch-to-host application would enhance or besmirch the corporate image. (Yes, failure to do this could prove extremely embarrassing.)

When calling the complaints division, for instance, the caller usually expects to air her grievances to a live agent. Her resentment and frustration may only build if greeted by a VRU unit.

· LANs, minis or mainframes — size up your host solution. For smaller call centers, a local area network can serve as the entire host side of the solution. Clearly, the last few years of application development have shown that a LAN-based or client/server-based application gives you more flexibility when it comes to importing telephone functions to the workstation than you'd get with a mainframe.

Of course, if you have a mainframe or mini already in place, you'll probably work with this existing hardware. You can often use these hosts as central servers, connected to workstations via local area networks, combining the flexibility of a LAN with the processing power of a mainframe.

· Pin down the vendor on probable savings and goals. Before you even think of contacting a vendor, you should evaluate the time it takes to handle a given call. Only then will you know what your projected savings might be.

Bear in mind that most applications salespeople are just that — salespeople. Get past the vague promises and pin them down on how much will be saved. Demand detailed projections and scenarios. Ask to speak to a few happy customers — even among happy customers you may find some potential drawbacks of a particular system.

If you're fortunate enough to be running a regional monopoly — a utility or water works — you can call colleagues from other areas to discuss any open applications they may have implemented. If you run a mail order house, you may have less luck getting your competitors to divulge their secrets.

· Starting over or improving upon the existing order. Many applications can be integrated into an open CTI environment rather painlessly.

For instance, you may have an application that calls up customer profile information by having the agent key in the customer's social security number. Using ANI, the open application automatically summons the file to the agent's screen simply by replacing the social security number with the caller's home phone number.

Many open applications, like predictive dialing engines, are more efficient or economical if purchased as turnkey applications. Sometimes, it pays to scrap the old order rather than undergo an ill-fitting adaptation.

· Test the waters. There are two ways to test computer telephony apps prior to full implementation. Dummy applications are available, simulating call traffic, your workforce, the equipment you plan to employ, your network services and your application code. You can also have a test region on the host, where you can run pilot tests while you're making changes, perform load analysis.

Many telecom managers prefer to phase in the new regime gradually through such separate testing areas. One way is to phase in with 10% or 20% of your customer base, then gradually broaden the application to include the entire base through the call center.

· Avoid glitz for its own sake. CTI apps perform some feats so stunning that even the most sober telecom center manager can get carried away with the fancy.

Few would argue that the act of automatically shunting a caller's vital data and his phone call to an agent's terminal before that agent even picks up can save a lot of valuable seconds in WATS time and agent labor. All of these precious seconds are lost, however, when the agent picks up the phone and exclaims — "Hello Mr. Brooks, how may I help you?"

If you call people by name before you give them a chance to introduce themselves, you're going to waste 20 seconds of your time with 'how did you know I was calling?'" The result is a transaction three times as long, and three times as expensive, than the manual solution.

· Don't ignore agent considerations. Weeks before you implement the application, you should set up a training program — coordinated by the applications developer — to master the system.

As with any introduction of automation, you may need fewer employees on the job — in this case, agents at their terminals. Perhaps your budget will permit you to divert customer service reps to a larger support group or complaint division. You may also be compelled to reduce the workforce, either through attrition or layoffs.

Also consider the implementation of new evaluation criteria for those who remain at the call center. If your application incorporates a voice response unit, for example, the unit will handle most of the simple inquiries without any live agent intervention. Which means that your agents handle only trickier, more difficult calls. So you should expect that the duration of an average call fielded by an agent will increase.

· Reality checks. Three months after your application is in place, then three months after that, you should take a look at how much you are saving.

Note that while it is relatively easy to calculate lower toll free usage or fewer agents needed to staff the phones, other benefits are more difficult to gauge — such as how many new policies have been taken out by insurance customers simply because the agent was able to transfer both their datafile and screen immediately from the life insurance division to the accident group.

Often, you'll find you must alter your long-distance contract, your agent scheduling, even the capacity of your computer plant to accommodate the changed call processing environment.

From a bottom line standpoint, though, these changes are probably for the better. Many end users report a nine to 16 month payback on their investment.

■ Key Features To Look For
And so here are five key features that you might want to add to your call

center through computer telephony. There are many others, and the list grows all the time. (That's part of the wonder of this industry.) Again, these are the ones I see most frequently, that provide demonstrable benefit to the companies that implement them. You can add them by adding software that contains them, or as part of a general CTI overhaul. The benefits of each of them are always the same — faster and better customer service.

1. Simultaneous Screen Transfer. An agent is speaking to a client. The agent has the client's database up on his screen. The agent needs to send the call to someone else for special treatment. Push a button on his screen, "Who would you like to send this call to?" He types the name, hits Enter. The call and the updated screen go to the specialist.

2. ANI/Caller ID Database Lookup. A call comes in. It carries the calling number. Your ACD grabs the calling number, passes it to your database over your LAN. As the phone rings on an agent's desk, the agent's screen pops with a screenful of information on the caller. What he bought last time. What his problems were. How they were resolved. What he tried to buy last time.

Automatic phone lookup can shave 15 minutes off a typical call — the time the agent takes to ask the hapless caller such questions as "What's your name, address, phone number, etc.?"

QUICK TIPS

1. Reducing headcount through shorter call lengths is a poor way to justify a CTI investment because there are many less complex ways to reduce call lengths and headcount is often not reduced by the amount calculated ahead of time.

2. Make sure you relate CTI to your company's three to five year business objectives, how you differentiate yourself from your competition and your customer service strategy. By doing this, your case to justify the CTI investment becomes all the more compelling.

3. Don't call in the vendors until you have done a thorough analysis of user requirements. Some vendors are happy to help you do this, but you may wind up with a set of requirements based on a vendor's capabilities rather than your needs.

3. Predictive Dialing. When your agents are not answering incoming calls, they could be making outgoing calls. "Last month and the month before, Mr. Smith, you bought four dozen boxes of paperclips. May we send you another four dozen?"

4. Other Database Lookups. Many agents are linked to only one database. But customers always want more information than one database can provide. A price list. A list of your dealers. Other machines you're compatible with. How to get the machine fixed. Layouts of the hotel rooms you're renting.

Voice Processing Fundamentals

Customers demand convenience. They want information quickly, but they also want specialized attention. And they want to reach you on their timetable, not necessarily yours.

Voice response assures callers reach the right department without the need for an agent. Callers like having options. They hate being forced to wait in queue. Voice processing means you can offer them options. Depending on the technology you use, they can leave a message for a return phone call, retrieve information themselves, or request that it be sent to them.

And the benefits to you are even greater. When you use a voice processing system, more calls get handled through the system. Instead of paying your reps to answer every call, they can handle just the callers who ask to speak to them. Depending on circumstances, you'll be able to handle higher call volumes with the same number of reps.

Information that an IVR system captures is always accurate. It comes firsthand, from the customer. By now everyone realizes the value of customer information. You can use it for cross marketing, surveying demographics about who your customers are, and so much more.

A lot of information about IVR will be presented in the next section. IVR is a special animal; it's the key voice processing component in call centers, worthy of special attention. What this chapter will do is explain some of the other voice technologies that are available. These technologies, like speech recognition and automated attendants are, if not critical, then important for specific applications and industries.

Less than 10 years ago it was possible to go through each voice processing technology and give an example of a standalone system that offered that technology. Today's systems are much more sophisticated.

These days certain technologies are found almost exclusively as functions in larger systems. It's likely that some or all of these are included in the ACD you've already got, whether you use them or not. (Maybe in reading this brief chapter you will see the virtue of simple voice functions and trot out your ACD manual to get some of them turned on.)

When there is a standalone product, it is almost always aimed at the low end of the market. But today's voice processing market is also a place where you can get what you want — exactly what you want. The hottest technologies are application generation software products, voice boards and the accessories needed to create "do it yourself" voice processing systems.

Here, I've outlined the technologies that are available. In a sense, all I'm really doing is showing you how some core voice technology (voice boards plus some software logic) can be put to use. There is precious little difference between auto attendants and their grown-up cousin, IVR, besides power, scalability and feature set. Under the hood, they are all essentially the same. But no matter what type of voice system you choose, they all share one common characteristic — you'll be more productive and will save money in the long run.

▓ Announcement Systems And Messages On Hold

The most basic building block in the suite of technologies called voice processing is the announcer. An announcer simply answers an incoming telephone call and plays a recorded message.

Digital announcers use a computer chip to store the recorded message. Other systems use tape to store the message, similar to the way an answering machine does. (Word of advice: stick to digital. Tape is too delicate, too cumbersome, and hard to edit. Digital is not just the future; it's the present.)

You can have the system play a message and simply hang up, or ring the caller through to your phone system after playing the message if they choose to stay on the line for more information.

Announcers can also work with ACDs to play messages to callers in queue. You can program a message to simply thank the caller for holding, play on-hold music, or even better, play recorded promotional messages.

Because an announcer is so simple, it doesn't have the high-tech appeal of other voice processing technologies. But announcers are vital to most call centers and many other businesses because they play music and messages to callers waiting on hold or in queue for a call center agent.

Call centers turn to sophisticated technologies like computer integration to save a few seconds per call — and may spend hundreds of thousands of dollars to do so. But few stop to think that a simple announcement on hold that tells callers to have a credit card ready can save that same call center five seconds per call at almost no cost.

When choosing an announcer the most important thing you'll need to decide is the amount of recording time you'll need.

▓ Automated Attendant

Automated attendants answer a call, play a message with a menu of options and route the caller to the extension or menu choice selected.

In call centers, automated attendants are helpful in having callers direct themselves to an appropriate queue. For example, an automated attendant in a call center might ask the caller to dial one for sales, two for billing or account information and three for technical support. The call would then be routed to the correct department or the correct ACD gate.

Once it was common to find standalone automated attendants. Today they are usually a part of a voice mail or other voice processing system. In fact, they are so fully integrated into today's PBX systems and messaging technologies that you almost never have to worry about buying this (or any of the voice technologies itemized so far). These are *features* of larger scale systems that are still important to the functions of a center.

■ Voice Mail

Now this is where you get into specialized technology. Not all voice mail systems are alike. And they have not (yet) been completely subsumed into larger boxes (though this is happening at a rapid clip).

A voice mail system answers telephone calls to individual phone numbers or phone system extensions, plays a greeting from the mail box owner and records the callers message.

At the mailbox owner's prompting it plays back messages, forwards them to other extensions, saves them or deletes them.

Voicemail has a different role in the call center than anywhere else. While it can be used in the traditional sense for call center managers or upper management, it's most often used to give callers an option to leave a voice mail message as opposed to waiting in queue for an agent when integrated with your ACD.

A voice mail system appropriate for the special needs of a call center should alert agents when there is a message in waiting. If the voice mail system you choose cannot do this, it's important to create a system designating certain agents to return voice mail calls when call volume falls below a pre-determined level.

Voice mail is a critical tool for the small center that cannot afford to staff agents after-hours. A voice mail system won't shut out any callers. It keeps your center open 24 hours a day.

Not long ago voice mail was the classic voice processing application and usually came in a standalone system that was sometimes bundled with an automated attendant. Today voice mail is usually a part of a complete voice processing system.

The latest thing in voice mail is "screen-based" voice mail that lets you call up messages of many kinds on your computer screen including your voice mail, email and fax messages. More and more vendors are jumping on the bandwagon to offer computer/telephony interfaces that put voice mail on your desktop PC. With this kind of interface you can get information not only about your voice mail messages, but also view email messages or faxes from your PC. This "unified messaging" gives you one mailbox combining voice, fax and email messages. Unified messaging gives you an easier interface than the telephone keypad. Most unified messaging runs across a LAN and integrates with your phone system. The object is to have desktop control over all of your messages, with the ability to retrieve them, read email messages or listen to voice mail and store and forward them through your computer or phone.

Some benefits:

· Users can view messages that come in even if they are on the phone.

· There's no need for dedicated voice mail hardware.

· The user can do all the configuring.

· You can move voice, data and email from site to site across networks.

· It eliminates the need to use the phone pad to issue commands.

Previously, telephone-oriented software had to run on systems connected to ISDN or proprietary digital phone lines to control features like hold, transfer and conferencing through TAPI.

With a unified messaging app, when a call comes in, a window on your PC pops up and gives you information about the call. With the click of the mouse you can ask callers to identify themselves, hold, play a greeting, transfer the call to another extension or ask the caller to leave a message without picking up the receiver.

As call centers change, combining different types of messaging will be more important, we will probably see the next generation of unified messaging applications focus more on serving the needs of the center.

■ Application Development Software

Getting a voice processing system to do exactly what you want can be frustrating. That's why call center managers with computer expertise sometimes create their own systems using a PC, voice processing boards and application development software (also called an application development generator, or *app gen.*

These software packages make putting together a system easier by protecting you from the lower level computer languages (read: "harder to use") through graphical user interfaces and object-oriented programming.

Sometimes a voice processing system will come bundled with an application development software component, to help you tweak the system to fit your exact needs.

When should you look into using application development software? If you are frequently going to modify your voice processing application — say for each campaign — then you could easily benefit from software that will let you do this yourself instead of waiting for your vendor.

But beware: this is not for the faint of heart. Even the most user-friendly app gen can be a beast. Quite frankly, creating the workflow logic of a voice processing application is very different from knowing how to schedule agents, or manipulate service levels. It's not always going to work out the way you planned it, and it's always going to take longer than you think.

Most voice processing systems come with preconstructed apps in their toolkits. If it's at all possible, try to use those. App gens are wonderful tools, but the closer you get to off-the-shelf, the better off you'll be.

■ Speech Recognition

Speech recognition is a lot like IVR, only callers get to speak selections rather than press corresponding numbers on their phone pads to get information.

Speech recognition gives callers without touch tone dialing the same access to information as those with touch tone service. Not only will it satisfy these callers — but think of the population of callers who need glasses to dial. These callers won't have to juggle their glasses with the phone pad to see the numbers they are pressing.

Although over-the-phone speech recognition still has a limited vocabulary, most systems are effective enough to allow callers to speak selections such as "sales," "flight number 123," "transfer cash" or "order baseball cap."

Speech recognition technology is constantly improving. Vocabularies keep growing (which means you can program the system to understand more caller commands). It seems that almost all systems are now continuous speech.

Make sure you choose one that is indeed continuous speech. Otherwise callers will be forced to pause and wait for a beep after saying every word or number. Since it's unnatural to speak this way, callers may be more likely to hang up or ask for a rep. There's also an increased chance of the system not understanding every word, since it's hard to tell speech from silence.

If you already own an IVR system and want to add speech recognition capabilities, you should check with your vendor. Many of the big manufacturers like Lucent, Syntellect and InterVoice have added speech recognition to their IVR systems.

This technology has made tremendous strides in the last few years. It promises to change the way customers interact with automated systems, broadening the range of telephony interactions.

There are two distinct kinds of speech recognition, known as speaker-dependent and speaker-independent. The two diverge wildly in the kinds of things they are good at, and the kinds of systems needed to make them run.

Call center apps necessarily focus on speaker-independent recognition. Many people will call, obviously. The human brain in the form of a receptionist can recognize a huge number of variations of the same basic input — there are literally an infinite number of ways to intone the word "hello." What you want in a call center is a system that will respond to the likely inputs — the most common words like yes, no, stop, help, operator, etc., the digits, the letters of the alphabet, and so on.

Internationally, touch tone penetration is still very low, leaving a vast installed base of potential callers who can not access IVR. It follows that these callers are then going to be expensive to process when they come into a call center because they have to be held in queue until there's an agent ready for them — high telecom charges from the longer than average wait, coupled with the cost of agent-service (rather than self-service).

On the downside, international call centers, particularly those that serve multiple countries, can field calls in multiple languages. If you use an IVR front-end to have the caller select their language then you by definition don't need speech rec. These are surmountable problems that have more to do with the operation of speech rec in practice than with the underlying technology.

Part Three

The Front End

Centers are moving away from the traditional voice-centric environments to multimedia customer contact centers that can process "calls" regardless of their point of origin (Web, phone or fax) and regardless of their form (voice, text or image).

Interactive Voice Response

If the call center is the front door to your company, then interactive voice response is the doorbell.

Simply put, interactive voice response (or IVR, as it's more widely known) is an automated system for collecting information from callers. It's a customer-oriented front-end for your call center. That is, it's a computer system that lets callers enter information in response to questions, either through a telephone keypad or the spoken word. The caller then gets some kind of information back from the system through a recorded (and digitized) voice or a synthesized voice. They can also get that information back through a connected fax system, or a website. How the information is delivered is less important than the fact that the customer can arrive, input and review data at any time, even when your center is unstaffed.

Whatever you can do with a computer, you can do with IVR. Customers can retrieve virtually any kind of data — from account balances to the weather in Chicago to the location of the nearest movie theater.

The benefits are vast. The telephone is familiar to everyone. It already has a worldwide network. Accessing information by telephone lets anyone interact with the computer from anywhere in the world. It also cuts down on the need for agents — especially when repetitive questions and answers are involved. Not only do you save on personnel costs, but also you are more likely to keep the agents you like, because their job is less boring.

Used as a front-end for an ACD, an IVR system can ask questions (such as, "what's your product serial code?") that help routing and enable more intelligent and informed call processing (by people or automatic systems). IVR far supersedes more rudimentary technologies (such as Caller ID) in such applications.

At one time there were no choices in how to implement it. You bought a dedicated box, integrated it with your ACD and the vendor would work with you to design your applications. Eventually you'd be up and running. So much has changed since those not-so-good old days.

The benefits today? Well for one, the market is truly open. For the most part, any ACD can integrate with any IVR system.

There are tools you can buy that let you design the system you want, called application generators, or app gens. These tools are simply software requiring industry-standard boards. Using these tools eliminates or greatly reduces reliance on IVR vendors. Such reliance (and it could get quite costly) used to be the only way to get up and running or to make program changes — unless you had knowledgeable well-paid programmers working for you. Now set-up has become inexpensive and almost simplistic.

When application generators first came out, the programming had to be done in DOS. Now everything is Windows, and development is graphical, so you don't need to be a programmer or telephony expert.

Even if you buy a ready-made standalone system, many vendors have developed enhanced, easy-to-use developing tools (such as GUI voice editors) to make it simpler than ever to be up and running, or to make changes to the program on the fly.

■ Why You Should Use IVR

This is a no-brainer. This is perhaps the most important peripheral technology you can implement. Depending on the application, you can siphon off from 20% to as much as 60% of your calls, or even more. There's no part of the customer interaction that *isn't* helped by the interactivity. The whole thing operates more smoothly because you've obtained some facilitating information from the caller. Sometimes that's nothing more than just a call's purpose — sales or service, Windows or Mac.

If you can get them to input their account number, you are armed with a database extract, and instantly you've created a CTI application that can bring rich screens full of information to the agent desktop.

Or you can deliver that information directly to the caller, no agent necessary, and improve your cost structure even more. Call centers are constantly caught between two contradictory imperatives: reduce costs and improve service. IVR is like a gift from the gods for both sides of that equation.

The fact that IVR is bundled with (or easily added on to) almost every switch sold in the last five years should make it an even easier decision. But if you still need convincing, here are a few thoughts.

IVR is widely accepted by callers. As the technology has matured, there has been more emphasis on correct implementation. Like eliminating "voice mail jail," that annoying state wherein callers can't press zero and exit the system. When badly designed scripts are eliminated, callers like the result.

IVR keeps your call center open 24 hours a day. Start from the assumption that you're never going to be able to hire as many staff as you'd like. Training costs are high, and with turnover high also, you've always got to have inexperienced reps in the pipeline. IVR can take the sting out of that process by giving callers an option during periods of unexpected high volume. And it's effective. IVR gives caller a nice impression of your company when they can get information at 2 a.m. IVR never calls in sick, takes vacations, breaks or lunch.

You can also use an IVR system to make handling calls easier for agents. One vendor's IVR system lets callers listen to company offerings, and when they hear something they would like to order or book, they just say "agent" and the system connects them to a live agent. Agents then hear an excerpt of what the caller expressed interest in so they know how to handle the call.

IVR can increase call volume. In one application over a six month period the call volume handled increased more than five times (from 30,000 to over 150,000 calls per month) At the same time, the number of messages left for agents remained relatively constant.

As voice processing systems take over more calls, the pie grows as well. These systems seem to create their own traffic. This causes problems for businesses that try to cost-justify IVR purchases through the number of staff positions they will be able to cut, because as an effective application creates new and better service, people will use more of it.

It's important to remember, with all this talk about cost savings and being open 24/7, that IVR can't replace humans, it can only assist them. In a well-designed IVR implementation, you've automated the transactions that can be easily done, leaving call center agents with the calls that require human skill and intellect. Some companies have gotten into trouble by trying to handle all calls through technology. The experts say that this is just not possible. Some calls will always require the kind of assistance that only a human being can offer.

Don't think of IVR as an all-or-nothing situation. Just because a system can do something doesn't mean you have to use it all the time.

IVR is cost-effective. Call center managers tell us an IVR system pays for itself in less than a year. That makes sense when you consider the number of questions it can answer without the aid of a rep. Most of the questions it answers are routine inquiries that would eat up valuable agent time. With IVR, agents can be left to handle only the more complicated questions or the callers who request a live agent.

What also makes IVR cost-effective is its flexibility. With most systems you can start small and just add lines as call volume grows. It's not uncommon, for example, to find that systems are bundled with pre-built software "templates" that you can snap together to create applications which you can then update when your needs change.

It gives callers control. Callers have options with IVR. They can press a key to reach the department they need, hold for a live agent, or get what they need without having to speak to a rep. A note of caution, though: during working hours, make sure callers can get out of the IVR system to reach a live operator. It's nothing but frustrating to press 0 trying to get a live voice and instead being disconnected or forced to listen to voice prompts again.

Not having a live operator available is only acceptable during after hours. In such a case voice prompts should announce something like "Thank you for calling XYZ company. While no one can take your call during these hours, you can press one to hear information on A, two to get information on B faxed," and so forth. A final prompt should announce "Press 5 if you would like to leave a message for a return phone call during office hours. Thank you for calling."

You can create special temporary applications. You can tie your IVR system into advertising campaigns. Each month you can offer different specials. Run an advertisement on television, in a newspaper or even in the yellow pages. Then, when callers reach the IVR system you can have a menu choice directly related to your ad.

When the special ends, you can return voice prompts back to normal. The benefit is that a large volume of callers gets fast, consistent information without bombarding your agents.

What To Think Of When You Shop

Evaluating various systems? Here are some selection and installation tips to keep in mind when looking over the vendor brochures.

1. Choose a system that lets you easily add more telephone interfaces and voice storage capacity — you should always anticipate growth. Line capacity describes the number of simultaneous conversations the system can handle. This requirement is a function of anticipated traffic, peak volume demands and the tolerance of the caller receiving a busy signal or a ringback of more than two or three rings.

2. User interfaces are typically subjectively evaluated during the system selection process, and are a function of the script and recordings. Recordings are usually first created by the installer, but updates are maintained via recordings made after the installation. Thus, the ease with which the system administrator can manage recordings is critical.

   The product should allow high quality recordings to be made directly with a microphone or telephone set, but should also support recordings made by commercial studios.

3. System usage reports are critical in preventing a business using an IVR system from isolating itself from its callers. The system must be capable of supplying informative reports about the nature and disposition of incoming calls, such as:

· How long did people stay on the line?

· How many hung up without making any selections?

· What items were selected most often?

· How many after-hours callers left messages for an agent to return their call?

4. Make sure the vendor understands exactly what you want. Tell them exactly what you want your customers to hear. Check references so you'll know their history of service and support.

5. Your system should not force regular callers to listen to lengthy prompts. Callers should be able to bypass recordings and skip to the prompt they want to hear.

Increasingly, thanks to better integration between IVR and the switch, you can offer IVR as one option to callers languishing in the hold queue.

■ Working With Outsiders

An alternative to rolling out your own IVR is to work with a developer. In some cases third party developers buy an application generator and develop an application specific to a particular industry. You can also hire them to develop the application you want.

Here are some reasons to use an outsourcer:

You can start small. If your call volumes are low, the vendor can start you off with just a few mailboxes or some combination of voice messaging and call routing.

Your current system has a longer life. Assuming the vendor's products integrate with your existing PBX (here I'm talking about very small centers), PC-based services and other hardware and software, you can enhance and expand your current system without starting over.

You are protected from obsolescence. Take advantage of state-of-the-art technology without locking yourself into systems that may or may not be appropriate in the future. It is the vendor's responsibility, not yours, to upgrade systems as new technologies become available.

Adding On More Features

IVR rarely stands by itself — it's the perfect integration technology for a whole host of voice and data applications. Add-on features like speech recognition, text-to-speech and fax-on-demand are becoming popular. It's important to choose a system with open integration so you can always add these features.

In narrowing your search down to the best vendor, look at the application you need, the services the company provides for support, and the image and reputation of the company.

Adding speech recognition is a time saver — especially when you have several prompts. A caller who knows what he wants to do can just say "claims department" and be connected. And as the Internet takes on a stronger role as a call center front-end, more and more companies will be offering access through all kinds of call center/computer telephony products. Some IVR vendors have already started.

Perhaps recognizing that the speech front-end is going to be at least as important as the touch tone interface in years to come, speech and IVR vendors are making hasty arrangements to work together, at least in presenting their products in an integrated fashion to potential customers.

Intervoice-Brite is acting as VAR for Nuance's speech recognition engine, for example. Nuance's speech rec is good, and is used in some notable high-traffic applications. Brite sells into some of the same markets, with large-scale IVR systems for government, financial services and telecom companies.

Combining with Nuance allows them to offer the base platform for front-ending the call center without having to develop speech rec on their own. It gives their customers the ability to choose speech rec as one option in front of the center, among others, that can include telephony input, Web and other CTI-enabled connections.

IVR companies tend to be very good at connecting with other companies for complementary product offerings — recall that several years ago they were among the first call center vendors to start building Web and email hooks into their systems, at the application generator level. Not so that they could start selling Web apps, mind you, but so that their customers could build them for themselves, or connect third-party service systems to the end user.

QUICK TIPS

1. Know Your Callers. To develop the best caller interface, look at the most common questions, comments and information your callers request. This should guide you in determining the types of inquiries you should let your IVR system handle. Getting a good handle on who your customers are and the reasons why they call can lead you toward an IVR system with applications specific to your needs.

2. Use IVR as a way to handle an especially large volume of calls for special applications. If Monday morning is your busiest time, rather than adding staff, use IVR to handle the extra calls. Or, instead of staffing up for special promotions and offers that you know will heat up the phone lines, consider ways you can use IVR to handle the callers who don't need to speak to a live agent.

3. You should not overuse IVR or overprogram voice prompts. Think of the application as a tree with branches. Too many prompts at once will confuse callers, or by the time they get to "press 6 for X" they will have forgotten what one, two and three announced. Three or four prompts is enough. After callers press a corresponding number you can have another three or four menu prompts lead to more options based on their first selection.

4. Always make sure that during business hours callers can press 0 to reach a live operator. They are, mostly, not stupid. They will try 0, and *, and #, and even pretend to have a rotary phone to avoid standing in the queue. Don't treat them like they are a mass of cattle to be herded towards your destination. The call costs you next to nothing, until an agent gets involved. Give them something to do, some knowledge of what they have access to, and what they'll need to provide when they finally do get to talk to someone.

5. Don't tell them to enter their account number or other information through the keypad unless you really, really intend to use it. Nothing makes people angrier, and makes you look more incompetent, than to have your agent ask them for something they just entered.

Chapter Ten

Speech Recognition

With almost no fanfare, speech recognition technology has made tremendous strides in the last few years. It's what you might call a stealth technology — the kind that keep academics and serious R&D departments busy for years showing incremental improvement, and then all at once the development reaches critical mass and it's everywhere, in all sorts of applications. It promises to change the way customers interact with automated systems, broadening the range of telephony interactions, and giving the call center a strong new tool on the front-end for capturing customer data.

The reason it is so explosive is twofold. First, the speed and power of the typical PC grew along the expected curve until it was strong enough to process speech in real-time. Second, the developed algorithms were steadily improved to allow computers to discern the appropriate patterns that underlie speech, without regard to accent, speed of speech or other eccentricity.

Speech rec is starting to gain a toehold in call centers as an autoselector — a tool that the customer uses to interact with an automated system to either route himself to the proper person (an auto attendant or ACD front-end) or extract the information he needs from a host database, à la IVR.

In the short and medium term, the interaction of choice for a customer wanting information is still going to be the telephone. While they are migrating to the Internet in huge numbers, call centers will still be deluged with phone requests for information, service, problem solving and order taking. IVR is still the dominant way for callers to routes themselves to their information destinations. When you put an intelligent speech engine in front of that you decrease the chances that the customer will ultimately have to make use of an agent's time. Costs are shaved and customers go away slightly more satisfied.

By itself, speech rec doesn't add new functionality to the call center. Instead, it adds new callers: those with rotary phones, those who are mobile, those who are so pressed for time that they can't be bothered to do anything but speak. It then processes those callers using the same traditional tools that call centers have used for years. The same benefits flow from speech recognition as from IVR: fewer calls that have to go to an agent, shorter calls, and more self-service.

This technology generates a lot of excitement in the public because of its association with things like voice typing, or dialing a cell phone by voice. But clearly, the specialty applications call centers need — those that need to be speaker-independent — are more powerful in the long term, with the potential to save agents time on data collection.

Consider an application created by Nuance Communications for Schwab's automated brokerage system. When I first saw this demoed in 1996, I thought it was pretty good: it understood me more than half the time, and seemed flexible. Now it's even better. And according to Nuance, it now handles half of Schwab's daily telephone stock quote volume, with 97% accuracy. With the migration of personal financial services to the Internet (and with price and service the determining factor in a competitive industry), giving a customer the ability to say "I'd like a quote on IBM" instead of typing out some ridiculous code is a key differentiator.

There are a lot of companies working on applications for this. As processing power improves and the cost of delivering a working application drops, it is likely that speech rec will take over as a successor to IVR as the "non-agent" telephony transaction.

The kinds of input that a speech rec system would have to process are very well defined-sequences of digits for things like account numbers, phone numbers, social security IDs or passwords. Or, some apps use discrete letters for getting stock quotes. There are a million ways to use it to extract information.

There are two distinct kinds of speech recognition, known as speaker-dependent and speaker-independent. The two diverge wildly in the kinds of things they are good at, and the kinds of systems needed to make them run.

Call center apps necessarily focus on speaker-independent recognition. Many people will call, obviously. The human brain in the form of a receptionist can recognize a huge number of variations of the same basic input—there are literally an infinite number of ways to intonate the word "hello." What you want in a call center is a system that will respond to the likely inputs—the most common words like yes, no, stop, help, operator, etc., the digits, the letters of the alphabet, and so on.

Telecom has gradually been accepting the technology in operator assistance and routing systems. (But not everywhere you think. Some automated applications that ask users for spoken input, like directory assistance, are actually just recording it and playing it for the operator, who inputs it manually—it saves time, but speech rec it isn't.)

Internationally, touch tone penetration is still very low, leaving a vast installed base of potential callers who can not access IVR. It follows that these callers are then going to be expensive to process when they come into a call center because they have to be held in queue until there's an agent ready for them — high telecom charges from the longer than average wait, coupled with the cost of agent-service (rather than self-service). On the down side, international call centers, particularly those that serve multiple countries, can field calls in multiple languages. If you use an IVR front-end to have the caller select their language then you, by definition, don't need speech rec. These are surmountable problems that have more to do with the operation of speech rec in practice than with the underlying technology.

Speech rec costs a lot to develop and perfect, but once it's done, it's done forever. The cost of maintaining it is negligible, and it has little of the headaches involved in CTI or other "fancy" call center technologies. Once you tease meaning out of the speech, it becomes input like any other, just like information entered via the Web, DTMF or told to an agent.

That's the essence of speech recognition in the call center — it's a simple front-end, with albeit limited application. But that's what they said about IVR ten years ago, and look where we are today.

Giant retailer Sears has turned to speech recognition in a big way, implementing it in 750 of their retail stores nationwide as part of a program to redirect calls more efficiently. It's also hoped that this will help them re-deploy almost 3,000 people to other, more productive tasks.

The automated speech system, built by Nuance, is part of Sears' Central Call Taking initiative. About three-quarters of the calls that come in to each of the stores' general numbers will be handled by the system; for a total of 120,000 calls each day.

Callers, when prompted, will be able to say the name of the department they want to reach — "shoes," for example. The cost of automating the department transfer is much lower than the cost of having an operator do it. (Reps are available to assist in case a customer has trouble.) Sears will be able to use this system instead of hiring temps during the busy holiday season and other peak periods.

The system was piloted at some Sears stores during a recent holiday season. Development began the prior summer, a fairly quick turnaround for a system complex enough to stand in for 3,000 operators at 750 locations. Sears integrated the Nuance system into their existing IT infrastructure; a custom app uses the speech rec system's ODBC hooks to query a centralized Oracle database for individual department phone extensions, reducing call transfer time.

This is one of the biggest examples of speech rec being used in consumer apps outside financial services. When retailers take on a technology, that's a sure sign they feel comfortable with both the consumer acceptance and the technical sophistication of it.

Airlines have also been historic early adopters. Airlines have been out in front in pushing this technology as a way for their customers to get quick access to a wealth of information. American Airlines has one such system, which they call Dial-AA-Flight.

The airline's automated flight information system gives customers data on arrivals, departures and gates. According to American, it handles approximately 19 million customer calls a year.

AA is moving to add a speech front-end to the system, at first in a pilot program (no pun intended) that will take 10% of the traffic. This is not American's first use of speech rec. In 1999 they rolled out a similar service to their VIP customers.

(It's interesting that speech, which just a few years ago was ridiculously bad at speaker-independent recognition, is now moving into real world applications from the top down, from best customers into the general pool.)

Financial Services Out In Front

Visa International is betting that "v-commerce," the heinously-named catch-all term for telephone transactions enhanced by speech-recognition, is a big part of their future. (Thank goodness "v-commerce" is trademarked, or it might actually catch on.)

Since 1995, Visa International has been an investor in Nuance, and has participated in pilot programs to add automated speech to cardholder transactions — basic things like card activation, card replacement, and travel planning.

These are things that, like all good IVR apps, don't need an agent for the basic, introductory information gathering stage of the transaction. Only when things get more complicated, or the consumer gets confused and tries to bail on the automated system, does an agent really become necessary.

Now Visa and Nuance are working on developing apps for use by the member banks for customers to self-serve over the phone. It's got some important names from the call center field, but it doesn't have any speech companies other than Nuance. Nuance's speech rec is as good as anyone's, but it's not the only one, so the impact of an association like this is in the power of the end-users, like Visa, to push a particular technology.

Several v-commerce applications are currently available for Visa member banks including speech banking and automated bill payment.

Another really interesting real-world application of speech rec is the installation that went into Ameritrade (the brokerage firm) in early 2000. Ameritrade decided to take their existing (and recently installed) speech rec front-end and expand its capacity.

Ameritrade's system lets their brokerage customers check their accounts and act on their investment decisions via telephone using natural speech recognition. The speech-enabled system was introduced to Ameritrade customers on March 10, 2000, and handled more than 650,000 calls in its first nine trading days. That huge response was a major factor in Ameritrade's recent decision to increase the port capacity of its InterVoice-Brite call automation system. The expansion is planned for this month.

Since the speech-enabled system was implemented, Ameritrade's call completion volume has significantly increased. The system currently handles an average of more than 85,000 calls on trading days. More than 40% of callers are already opting to use the speech recognition capabilities. This self-service transaction option gives Ameritrade customers a faster, more convenient method for rapid stock transactions while enabling the company to increase the efficiency of its call center by reducing call wait times and freeing agents to process more complex customer service requests.

The system isn't pure-speech only; rather, their call flow involves an interesting hybrid of traditional user-entered touch tone digits and speech input. To use the self-service stock trading system, callers enter their account code and personal identification number, using traditional keypad entry. Then, the system asks for which stock the caller would like a transaction. Rather than using a tedious touch-tone entry method, callers speak the company name or stock symbol in a natural voice.

It runs on InterVoice-Brite's *OneVoice* platform and uses technology from SpeechWorks. The system's vocabulary exceeds 60,000 words and even recognizes popular stock nicknames, such as "Big Blue" for IBM. (Though I believe you'd have to be really wanting in common sense to actually try to trade stock using that kind of nickname.)

InterVoice-Brite has also put in speech-enabled stock systems for DMG & Partners Securities, Lim and Tan Securities, Keppel Securities in Singapore, and Hyundai Securities in Korea.

The system will be able to interpret more than 80% of first and last names in the United States. That's going to make the system more viable for applications like health insurance benefits verification, travel reservation verification and cancellation, and inquiry applications where callers are asked to leave their name and number for an informational callback.

New multilingual capabilities simultaneously support two or more languages on a single system. These multilingual capabilities will enable application developers to create self-service applications in 10 different languages. Callers can also respond to prompts in their own dialect depending on geographic location and regional demographics.

Languages supported now include English (US, UK, Australian and Singaporean), Spanish (Latin and US), French (European and Canadian), Chinese (Mandarin) and German. The system features a vocabulary with more than 70,000 words and supports an increased number of ports to support higher call volumes. Additional tools include custom vocabulary development, industry-specific grammar libraries and a self-tuning feature.

▓ Speech Rec: The Gateway to CRM?

A lot of attention has been focused lately on the pros and cons of CRM tools — much from the point of view of "which tool can I buy and what is it going to do for me?" One of the things that's been overlooked is that for CRM to function properly, it needs to receive constant, meaningful input from an outside system. Input that acts as the raw meat for its real work, which is the analysis of what the customer wants and whether he got it.

We're used to thinking of two main data inputs from the customer. One, IVR, has been around forever and is widely understood. There are not too many new ways to construct a branching tree routing script, or to parse an incoming account number. Application development for IVR has been made easy enough for an intelligent non-expert to do a credible job putting together working, useful apps.

The other input mode is the very rich, multi-textured Web interface. This has developed so recently that there are many techniques for feeding customer input from a website through to a CRM system and into a contact center. All the many data-gathering and connection modes fall into this camp: email, click-to-call, text chat, and Web-based ecommerce forms.

Almost off the radar, though, is another technology that has reached strong maturity. Speech recognition, which is making big waves out in the consumer technology world, is still seen as something of an afterthought in the CRM/contact center world. In fact, it's a lot more than just a replacement for touch-tone input.

Why is it different? I'd like to argue that it creates the opportunity for a completely different type of interaction than does IVR. IVR, despite having "interactive" as part of its name, is really a one-way channel. Customers enter ID and pull out a small subset of data that pertains to them. Remember why it caught on in the first place: it automated the dumping of small bits of repetitive info to the caller, keeping those calls away from expensive agents. Relatively low tech, profoundly efficient, easy to diagram into an existing call flow — the very definition of no-brainer.

But when you add speech to the same call, you add several orders of complexity. Forget about the complexity of the tech that you need to operate it; instead, concentrate on the complexity of the information flow back and forth between you and the customers. Instead of asking questions that get answered only in numeric digits, you can draw out responses that are far more nuanced and subtle. Few people are going to enter an address using keys that have three letters apiece. Asking for a stock quote using the Schwab IVR is hard enough — each letter of the alphabet is assigned a two key code. They had to send customers little wallet cards to remind them of the alphabetic cipher just to be able to retrieve stock quotes. This was in 1996, before they were heavily online, and before they installed a speech rec system.

Stock quotes are one of those really basic information retrieval apps that the Web does really well as a replacement for the phone system. But you can do things that are so much richer, limited only by the system resources available to parse the speech, and the power of the recognition engine.

SpeechWorks, one of the companies with powerful speech rec tools, says that to run high-quality speech applications, you need four things:

· state-of-the-art technology;

· high-level building blocks (essentially this means that the recognition engine contains prefabricated modules for handling certain types of speech);

· tight integration on robust telephony platforms; and

· tools for analyzing and tuning applications.

That's to make the speech rec work; to make the application it's running a success as well, you also need

· appropriate application development procedures;

· an understanding of what your app is ultimately supposed to do for you, in terms of what would make it a success; and

· a good user interface.

SpeechWorks says that based on several of their customer installs, the average cost of an agented call per minute is $1.50; by contrast the average cost of a speech-rec attended call is just $0.25-$0.35. That's not too surprising, and they rightly say that the speech-rec costs vary based on the underlying contract the call center has with its local telcos for long distance traffic.

At one of their installations, the length of customer interactions was reduced from 12.5 minutes through touch tone to two to three minutes using speech. This goes right to the question of speech rec as a rough equivalent to IVR. If anything like that reduction can be repeated across the board, or even in a significant minority of applications, then speech looks a lot better as a way into the database despite the higher level of technology it needs to implement.

From a call flow and design point of view, though, it would be a mistake to think of speech rec as "talking IVR." When I speak of a richer interaction, I mean this: you don't have to delineate options one through four and leave the person scratching his head to figure out where his particular problem fits into your schema. The sophisticated application will acknowledge that there are ambiguities of response, and will tailor prompts to try to zero in on what the customer needs without being as linear as IVR.

People who are expert at using the system can shortcut through it, for example, or can barge in (that is, talk while the system is talking and have it know that it should stop and listen).

Those points apply to the IVR/speech rec comparison, which is how you look at speech rec when its main purpose is to identify the person and route to the right agent. But again, the interaction can be richer, used to gather information that you didn't have already. Once you've used the system to identify the caller, you can ask questions that have more detailed answers, even questions that are tailored to a particular audience or context. The stronger the speech rec engine, the more you'll be able to parse out of what a caller says. Again, its strength in the long run is not going to be that it gets the information to the caller at a lower cost; rather, it's going to be that it gets information from the caller to you in a more meaningful and spontaneous way. It's easier to say something into a phone than it is to fill out a survey and mail it back, or even to fill out a form on a website.

When I look at the spectrum of CRM-style applications that are rolling out over the next two years (and it's a long list), the common element is the need for an information channel that brings information reliably from the customer inside the company. We're used to customers calling when they want something, and parsing the data that comes with the call is so old-hat it's never even mentioned anymore. We're quickly getting used to customers using email and Web for interactions.

In a conversation with a very smart Dictaphone executive recently, it was opined that we're moving inexorably to a point where all transactions are recorded, stored, and analyzed using advanced data mining techniques. He was speaking from the point of view of quality assurance and agent performance, as well as customer satisfaction measurement. It strikes me that if universal recording and archival does arrive, the collection of speech rec data, added to the analysis, could be a valuable (if a bit spooky) addition.

CRM, so much a buzzword now, is an idea that stands in for a range of future technologies, some of which will catch on and some of which won't. It's the theory that matters, that information flows freely between all systems and is parsed somewhere inside the organization, probably away from the call center. The theory of CRM will depend on controlling the most customer information at the lowest cost. Right now it seems that speech recognition has a good shot at replacing IVR as the primary information gathering tool for phone-only interactions.

The Web & Call Centers

When the first edition of this book was published, not that many years ago, the idea of a connection between the Internet and call centers was theoretical, and not very well thought out. The most common way of thinking about the issue was to imagine self-support applications: literally, email-based posting of customer support cases, and the ability of a customer to search a database of problems and solutions by themselves.

Just a few years ago, it was possible to talk about the different ways of delivering customer service without mentioning online support at all. When you did mention it, it was in the context of CompuServe forums or company-sponsored bulletin board systems. The Internet explosion that's brought us webpages was still a year or so in the future. That's how fast things have changed.

A call center is not a place. It is a set of functions. It is the process of selling to people who are not in the room with you. And of serving their many varied needs. A call center's primary function is to create and keep customers.

Right now, that function requires a physical presence, a location — an actual set of seats filled with people to help customers. Agents that have access to stores of data about the customers and the company, and the points of intersection. These people act as gatekeepers for the two-directional flow of information — as intermediaries and interpreters.

What if those functions could be accomplished without people? Or with a vastly fewer number of people, so that those who are left are true experts who add value to the transaction, and who don't merely funnel that transaction along.

This is the goal experts are reaching toward when they tout the Web as a useful adjunct to the call center — the promise of a workforce deployed in exactly the way that is most useful and efficient. Customers who solve their own problems, who in essence sell themselves, and a specially trained cadre of agents dedicated to doing what only humans can do.

To some extent, the call center industry has been flirting with this notion for years, with varying degrees of success. First there was fax, then fax-on-demand. Want information about our company at 2 am? Rather than pay to staff off-hours, make information available for retrieval by the customer himself. It's fast, cheap and gets generally high satisfaction marks from the customers.

On the other hand, some of the lessons learned by newer ecommerce and etailing (hate those words) companies haven't yet been taken to heart. As a benchmarking survey commissioned by Swallow Information Systems has revealed, organizations are turning a blind eye to customer complaints and not treating inquiries seriously.

The survey found that 93% of companies in the business-to-consumer marketplace gather customer data, but only a third convert opinions into improved customer-led policies. 89% use the Internet as their prime way of contacting customers, but most have no procedure for resolving online complaints and inquiries, and over 25% have no dedicated customer service policy at all.

The survey was conducted across 28 companies, including SOCAP (The Society of Consumer Affairs Professionals) members, blue chips and market leaders, including retail, telecom, travel & leisure, financial services, FMCG and manufacturing industries.

The research suggests that the problem is caused by a 'quick-fix' approach to complaint resolution, where staff members find a solution then move on to another task, rather than upgrading the service for every customer. Another key cause cited was a lack of awareness of the technologies available to receive and respond to customer contacts, and how customer information can be shared between departments.

"It's alarming that companies are underestimating the value of their customers to such a degree," said Ros Gardner, vice president at SOCAP UK. "The customer is an organization's most important resource, and they should be aware that listening to their opinions is one of the best indicators of success or failure, and where improvements should be made."

"Companies are projecting an image of good service, but too often customers are being ignored," said Bill Bostridge, VP of sales at Swallow Information Systems. "Providing a channel for customers to contact you is the first step to good service, but collecting data without listening to their opinions is a pointless exercise. Managing customer contacts is the key to turning around services that customers actually want and need."

Then there's IVR. It's great for routing calls to agents, shortening call times, getting people into the right queue, etc. But it also allows people to self-serve for simple database lookups like an account balance, an order confirmation or shipping status. Or to diagnose a technical problem.

But still, there is an unstoppable trend toward providing an automated response to customer interactions. The reasons are clear:

· Automated responses are cheaper than agent-provided ones.

· They are always the same for all callers. Two people who call for directions from the airport to your office won't get different routes from different reps.

· Automation is always available, even when you're closed.

Of course, there is a downside. Some people miss the personal touch. And any problem or question not planned for in the rules-based structures of your system requires human intervention anyway.

The Web is the third advance in self-serve automation. It has many of the advantages of fax — it's a dynamic, easy-to-maintain format. It is available to huge numbers of potential customers. And it surpasses fax or IVR in one critical area: it has enormous multimedia capabilities (sound, graphics, video, and more). It is rich in the one quality IVR lacks — the ability to control applications that require visual presentation or extensive keyboard output, or both.

Customers can order products. They can download software. They can read catalogs.

Clearly there is a need for alternate points of entry into the call center. You can think of the call center as the focal point of a "customer contact zone" where a lot of interactions take place. Ultimately, as more kinds of call center applications are developed that bypass the agent, a given customer will have more choices for entering the zone and concluding the interaction. Some points of entry may be better for making a sale — document retrieval by fax-back, for example. Others are better for customer support, like IVR.

The Web offers an amalgamation of these techniques. Where the Web first began to take hold was with help desks. These smaller centers, already strained by escalating call volumes, were in the vanguard of agent-enhancement technology. Problem resolution software frees technical experts from the drudgery of answering repetitive questions, letting them get to the business of solving more complex problems. It puts them in a position to add value to the customer transaction, rather than merely pipeline a piece of existing information to the customer.

But what about other call center systems like ACDs? If a caller has a choice of how to get into that customer contact zone, where will the switch fit in?

The answer has turned out to be twofold.

First, there is the omnipresent email. (Itself a revolution in communications technology; if it weren't for the fact that the Web has pictures and sounds, we would be marveling at the amazing transformation email has wrought in our society.) In call centers just the last two years has brought something that's sometimes called the "Internet ACD" or "Email ACD." Bad terms, but they do describe fairly well what's going on.

What they do is perform the same types of targeted routing that the telephony switch does for calls. They take large volumes of email as it comes in, parses out some form of meaning (who is the sender, what is the subject, etc.), determines whether it can be answered with an automated response, and if not, sends it along to someone who can handle it in an effective way. Routing tables show whom can handle what, and how often.

These systems also track and audit the response to those emails. So you can set an organizational parameter, for example, that all emails have to be "handled" within a certain timeframe.

The other major thrust has been a tentative exploration of one of the Web's major attractions, which is live chat. The idea being that a customer visiting your webpage would click on a button to initiate a text chat window, having a semi-real-time conversation with a rep back at the call center through a text typing session.

There are several advantages to this. One is that a single rep can handle multiple chat sessions at once, because many of the responses can be automated. (One vendor showed me a system where a rep could handle six at once — a recipe for burnout and turnover if ever there was one.) The other advantage is that it is relatively simple technology, that connects it to a rep's desk, in contrast to the much more complicated Web callback systems that attempt to connect a Web surfer to a call center through an actual telephony connection.

As you would expect, vendors are closely watching Internet technologies, looking for ways to integrate their switches with Web-enabled applications. As one manufacturer suggested, it is when the consumer can (and does) cherrypick from a combination of entry points — fax, email, Web or voice call — and expects to switch from one mode to another during a single "interaction" that advanced ACDs will need to be tightly integrated with the Internet.

What will happen is that human interaction will be reserved for where the agents can add the most value.

What We Are Dealing With

Buchanan Email, a British consultancy specializing in the use of email, performed a website response survey that attempted to assess the current standard of customer service by email in the UK. The Buchanan survey visited 361 website addresses across several industry sectors. They sent an email inquiry to each accessible website, asking general factual questions, taking into account the product or service offered by the individual sector or specific website. Then they allowed 28 days from the sending of their email for a response to be received. Of the 290 successfully delivered emails, the surveyors received a response from 180 websites within the 28 day period — a 62% response rate overall.

You can make allowances and say that the need to deal with high volumes of customer email runs ahead of a company's ability to do so, that it's always a surprise how many come in. But the simple fact is that customers are turning to email for queries at a much higher rate than anyone expected. Dealing with those emails should be a higher priority than it is. Maybe the call center isn't the right place to do it; maybe we'll see the development of dedicated email centers. In either case, it's plain from the Buchanan data that personal responses are, at this stage, better for the consumer than automated ones.

In any case, however, research seems to show that there has been a rapid move to integrate the Web into call center applications. According to a recent Frost & Sullivan report, the compound annual growth rate (CAGR) for this market for 1997-2004 is forecasted by them to be 110.4%.

They find that call centers are moving away from the traditional voice-centric environments to multimedia customer contact centers that can process "calls" regardless of their point of origin (Web, phone or fax) and regardless of their form (voice, text or image). This finding is not exactly news, but it's nice to have data to back up my strong anecdotal feeling about the way that the market is heading.

F&S also find that there has been a proliferation of companies entering the market since 1998, and this includes companies coming from both sides of the fence: Internet companies and call center companies. In many cases, these companies are partnering with each other to leverage expertise in each of these areas.

For the purposes of this study, a Web-enabled call center was defined as one in which there is a system that links a call center to a website and enables a Web browsing customer to interact real-time with a CSR. Two-way, real-time communications between the CSR and end user can be achieved using either callback, text-chat, voice-over-IP, video-over-IP, and whiteboarding.

What About Internet Telephony?

While it is true that live audio/video Internet products are out there, in the customer service environment (that is, in call centers), these products are almost never implemented. At the present time, Internet telephony is not a factor in this arena, and quality of service is just one of the many good reasons.

Web/call center combinations are being used in pilot applications, and often they incorporate a "call me" button on the Web page that brings the customer into the call center for a telephony interaction in parallel with what's going on over the Web. More often than not, these things initiate an outbound call to a customer-entered telephone number.

The problem implementing IP-telephony based service has less to do with the network's quality of service issues than it does with the way calls are routed into the center (the ACD has to be able to send it to the right agent, and queue it, and report on it) and with the human issues of how customers wish to get their service. In nearly all cases, the telephone is the instrument of choice for service delivery, despite the availability of a Web alternative.

One vendor, CosmoCom, is making a go of an IP-based virtual call center system that blends traditional voice telephone calls with live Internet sessions. It manages and distributes both live calls and messages, including voice, fax, and email.

CosmoCall supports fully distributed operation with remote agents and multiple site operation transparently through its IP backbone that transports voice and data. It includes a multiple chat capability that lets CSRs conduct several concurrent text-based chats, while simultaneously speaking with their customers over the telephone or the Internet.

With *CosmoCall*, visitors to an ecommerce website can click a link to establish a live multimedia connection to a customer service representative. *CosmoCall* queues the calls, selects the right kind of representative for each caller, and informs each representative about the nature and context of the call.

And on another front, Dialogic demonstrated an interface between Internet telephony and CTI applications. This gatekeeper-based technology extends the capabilities of Dialogic's *CT Connect* call control software. According to theory, applications will be able to monitor and control IP telephony calls in the same way they currently do in traditional telephone environments.

This new technology will allow CTI applications to operate within IP telephony environments as easily as a traditional PBX. This transparency between IP telephony and traditional PBXs means that CTI developers will be able to market their applications for use in IP telephony environments with little or no change.

Customers will then be able to use IP telephony in conjunction with contact management, customer service, and support applications: in other words, in call centers.

In a standards-based IP telephony environment, the gatekeeper is the focal point of enhanced call processing because endpoints within its domain consult it whenever a connection is set up, torn down, or changed. During call setup, for example, a gatekeeper handles functions such as bandwidth allocation, address translation, access permission, and call routing.

■ The Connections in Practice

A book is not the perfect place to talk about such a dynamic and fast-changing subject as the relationship between call centers and the Internet. In any case, one way to see the impact of the changes is through the flurry of products that have come out recently, and the changing relationships between companies as they cross from one product category to another.

For example, a Web-browser-based publishing tool makes collecting and sharing customer data within the call center, and throughout the enterprise, as easy as accessing a webpage. Envision Telephony's *SoundByte Enterprise Web Desktop* lets centers publish customer contact information in a browser page format or "Web desktop."

The new format simplifies the transfer of customer information from the call center to the enterprise. Departments outside of the call center can use *Web Desktop* to get customized access to data necessary to make more strategic, timely business decisions. The kinds of data at work here are primarily digital call recordings and call center performance reports.

Executive management, marketing and product development, for instance, can track customer response to promotions, monitor service quality and query customers through this system. *Web Desktop* is part of Envision's *SoundByte Enterprise* call center suite, an automated call monitoring system that collects and publishes information about customer contact with a company's agents.

Core elements of *Web Desktop* include call evaluations and graphical comparisons of individual versus group performance. Supervisors can also add training tools, demonstrate productivity reports, publish department specific issues, and highlight morale topics. For example, supervisors can feature "agent of the month", incentive programs, and other events.

VocalTec's *Surf&Call Center* lets online shoppers make toll free calls directly from a website into any telephone or formal call center. A *Pro* version goes on to let online shoppers speak over the Internet to call center agents as well as jointly surf webpages and fill out online forms in a secure environment.

Surf&Call Center uses the same standards-based IP telephony network infrastructure that carriers can use to target a variety of audiences with services such as phone-to-phone, fax-to-fax, PC-to-phone, and Internet Phone Call Waiting.

The SDK also allows system integrators to customize webpages and agent desktops to support *Surf&Call Center's* voice and joint data collaboration capabilities.

As previously stated, the market for Web-enabled call center solutions is expected to grow at a compound annual growth rate of more than 110% through 2004, according to Frost & Sullivan stats. Interestingly, this product is aimed squarely at the carrier market and the call centers those carriers spawn.

Primus's *WebPack* is a set of tools for creating knowledge bases and publishing them on the Internet, complete with natural language search capabilities. The gimmick is that you're supposed to be able to complete said knowledge base within a single workweek.

The advantage to putting the knowledge out in front of the call center is simple: customers will use it to try to solve their own problems *before* coming to an agent for help. This cuts down on call volume.

(Though I suspect that the duration of the remaining calls would be somewhat higher, because those calls are likely to be the thorny ones that can't be worked out using an automated system.)

One early user of the *WebPack* system reported that for 65% of its customers, the self-service option directly solved their problem.

Another Internet front-end product takes a somewhat different approach. FaceTime's *Message Exchange* service tries to meld all the newer modes of Internet communication — instant messaging, for example — into a logical package that can be parsed by a call center.

Consumers engaged in online shopping at a FaceTime-enabled website instantly receive live and/or email-based sales and customer service assistance by connecting to the company's call center. The *FME* service suite routes, queues, and reports on inbound email and instant messaging interactions, helping sites prioritize customer requests and manage service levels.

Most interesting is *Instant Messaging Management*, enabling ecommerce sites to manage real-time, live text-based communication between a CSR and a customer. FaceTime accepts, routes, and queues these requests to the appropriate CSR, and provides supervisory functions and management reporting. (A lot of the actual interaction can be automated.)

The other main component, *Email Management*, manages inbound or forms-based requests from consumers.

The FaceTime *Message Exchange* features a real-time monitor that shows supervisor and managers current queue and agent status. It can also monitor customer specified service level targets and alert the supervisor when these targets have been exceeded.

FaceTime is an application service provider (ASP) for Internet customer service. In other words, FaceTime is an online service, not a product. That's good, because it means you can be deployed in days, scale up or down and avoid implementation hassles with the IT department.

The Fallacy of Email

In just a few chapters, I'll get into a discussion of some of the interesting new tools available, including CRM and the new theories of multichannel access for customer contact. But it seems appropriate to pause here for a moment and discuss, since I'm on the subject of the changes wrought by the Internet, the strange nature of email in call centers.

With an astonishing speed, the way we discuss the relationship between the call center and the Internet has changed. It's not going to be about "click-to-call" buttons on Web pages. And it's not going to be Internet telephony or IP networks handling calls.

Those things will happen, but not so soon as the vendors of those technologies would have you believe. No, what's really happening is email, and its cousin the Instant Message, and its other cousin the live-text chat.

This makes a lot of sense, when you think about it. The call center was built on the premise that customers initiate contact, and that whatever they need can either be handled in real time by an agent, or handed off to an automated system that can extract the information needed and deliver it to the customer.

Email is just a phone call that happens in slow motion. It's trackable, contains its own audit trail, and can be parsed by computers looking for keywords. And most important, it's ubiquitous. Everyone has it. Not as many people as have telephones, certainly, but an explosive number.

If you run a call center, and you believe that the Internet is coming your way through a call-me button, then you have to make certain assumptions. You have to plan for how the webpage is going to connect to the call routing engine, for one. That's a complicated piece of business; you have to make certain assumptions about queuing that may or may not be accurate because people surfing the Web are not the same kind of randomness as people picking up the phone.

And you're going to have to deal with the complex CTI interactions between the forms that they fill out online and the data residing back at the host. Because it's silly to collect information and then not coordinate it back at the agent's desk. And even sillier not to take advantage of the connection to do screen sharing between the agent and the customer.

But once you've begun to travel down this road, you've introduced a lot of new costs, and managerial headaches. So many, in fact, that I've been able to find very few call centers that decided it was a road worth travelling — at this stage.

Many people believe that this kind of Web/call center integration is only a few years away, and I can see that logic. But it's not a here-today kind of technology, for most people.

But while this debate was going on, email kept coming. It's flooding into companies in much the same way telephone calls did in the early days of call centers. It's often undirected, (what intelligent Webmaster, after all, is going to answer 10,000 emails about why a product doesn't work?) and uncategorized. It's a mess. Just as the days when call centers were about triaging the incoming calls and finding someone — anyone — to handle the traffic.

The response to this phenomenon has been very interesting. First, a set of tools was developed to tag and track incoming email for companies. The first thing you have to do, obviously, is identify what each one is about and try to match it up to known data about customers. After that you apply some rule to decide what the appropriate response should be.

Lots of little companies entered the fray, many (if not most) from the Internet side, not the call center or telecom industry.

Then, as things developed, these small companies aligned themselves with the voice switching firms, and others that provided call center infrastructure. In a strange paradox, the call center has grown so important to corporate well-being, and has taken on the role of "customer contact center," that it is a natural place to push all these unsorted emails, whether or not the call center is set up to handle them.

The fallacy of email, however, is that handling an email is not the same as handling a call. It's similar, but that's about all you can say. Expecting a call center to do double duty as an email processing center is as hazardous as expecting a single agent to handle both inbound and outbound calls. Most centers don't do that, because they know that the skills required are different.

Some centers begin this process by assigning some agents to email processing part time. They can route email for response to a particular rep or group during preset low points in the call traffic, for example.

But this makes it hard to assess the agent's overall productivity, because he's doing different kinds of work that require different measurements of success.

And it prevents you from setting the same kind of consistent service level targets for email that you would for calls. You can't very well expect to respond to all emails within 24 hours (a common target) if you don't know at any given moment how many people will be available to handle emails, or how many you're going to receive.

If you truly believe that an email from a customer is as important — as valuable — as a call, then you have to treat it equally. And you can't rely on handling it during the call center's off-time.

What's true for email is also true for some of the other email-like interactions. Text chat is where the agent and the customer are connected through a text window and type back and forth at each other. Some products that facilitate this are proprietary, and others use one of the open Instant Messaging tools like those from Yahoo or AOL.

True story: last year I was watching a demo of one of the new text-chat systems. It was a fine piece of technology. The system worked as advertised, it seemed easy to use, and looked good from a customer point of view. It contained software to parse what the customer was saying and present the agent with a choice of responses to save time.

The exec at the company showing it off had six windows up on the screen, meaning that the agent was engaged in six simultaneous customer conversations. I thought this was just a bit of showing off, until he said that they were selling that point to potential customers, pointing out that they could have each agent do six times as much work.

Think about this: in a world where 35% turnover is not unusual, does anyone expect a call center rep to carry on six customer conversations at once? When not on the phone? You'd be crazy to pitch that to reps.

I'd be surprised if they didn't balk at two.

What this argues for is a separation of function. Email, chat and other textual communication should be in the hands of people specially trained for the job. There are nuances to it. There are special skills that should be identified as key to this kind of work. Training should be dedicated to it.

Maybe there are people who can do both, but I don't think you'll find enough of them to fill all the call centers.

I think people who choose to go into the customer contact field — which is a better way to think of call center repping — should be tested for the kinds of interactions they excel at, and should be put on either email or chat or phones, depending on what suits them. Maybe some of each, to create a backup in case of emergency.

The fallacy of email is that although it has a lot of the same qualities as a call, it's a different kind of interaction. People have different expectations for what's going to be in the response. And there is no consensus on when they should expect that response.

The explosion of tools, many of them wonderful, from companies like Mustang, FaceTime, Kana, eGain, Brightware, WebLine and many others speak to the health of call centers. It means they are grappling for solutions to a growing problem. You could see it as a problem of riches: many customers using more channels means more information gathered about their likes and dislikes, more opportunities to make sales.

The problem is not with the tools. It's with the assumption that the person handling the calls is the right person for other kinds of connections, as well. Let's experiment with dedicated email groups. Don't build "email centers" just yet, but separate the groups who are best able to handle emails. And don't jump to morph your call center into a "customer contact center" unless you're really willing to deal with the management responsibilities that come along with it.

Chapter Thirteen

Giving Video A Second Look
By Madeline Bodin

You probably haven't given serious thought to putting video in your call center. (No, not showing videos to your agents. They really don't need to see As Good As It Gets at work. I mean adding video call handling capabilities to your call center.) A year ago, when I wrote an article on multimedia call centers, even the call center technology vendors I spoke with weren't very enthusiastic about video in the call center. They were doing handsprings over Internet integration, but that is another story.

I thought video was overkill myself until I opened an IRA (individual retirement account). Now, I live in the country — way out among the cows and sheep. There are people nearby whom I'm sure are perfectly good financial consultants. But it just so happens that one of my best friends is a financial consultant. And it just so happens he lives 300 miles away from me.

He's such a good friend that he is licensed in my state, just so he could handle my account. He mailed me a huge packet of stuff — and then the fun began.

Filling out the paper work to open an IRA long distance is a multimedia experience. We sat on the phone with the papers in front of us. My friend tried to walk me through the forms, and some were easier than others. I had to fax stuff to him. He had to fax stuff to me. I had to mail stuff to him. He had to mail stuff back.

Most frustrating of all was not being able to find exactly what he was talking about instantly as we went through a 25-page report together. I got lost more than once.

"We need video," I told him. And since he was in financial consultant mode, he gave me the polite version of "Yeah, right."

I didn't need to see my friend's face as we filled out this form. He didn't need to see mine. That's the biggest misconception about what video can do for your call center. Moving pictures of your customer or your agent are not much help in anything (except perhaps a plastic surgery consultation) no matter what any of the other proponents of video in call centers tell you.

A video connection lets you transmit anything you can show on your computer screen, including any sort of form: IRA, catalog order, medical claim, home owner's insurance, auto loan, mortgage, college application.

Not only do both parties get to see the form, but they can also both change what is there. My friend could have circled the information he was talking about, and I wouldn't have been looking at figures on another page. He could have highlighted the part of the form I was supposed to fill out, and he could have seen that I was doing it correctly.

The faxing and mailing could have been kept to a minimum and the conversation would have gone faster and smoother. Video would have even let me transmit my signature on the form.

I would have liked to do all of this from home, but most video applications revolve around a video kiosk. (Think of something like an ATM, only you get to sit down. And yes, banks are among the first users.)

But, I don't want you to think that implementing video in your call center will be easy. It won't be for a few reasons. The paramount reason is that video in the call center is still an emerging technology. Things change daily. Vendors work with completely different standards and finding someone with experience making everything work can be difficult.

Among vendors, there tend to be switching experts and video experts. The smart ones work together, but there are many good solutions among all the options.

If you already use a sophisticated call center switching system, such as Lucent's *Definity* or Intecom's *E* platform, and you want your video agents to work from the call center, adding video to the call mix can be relatively simple.

"At Lucent we are applying all of our experience with routing and reporting to multiple access," says Laura DiSciullo of Lucent. "Video calls can be queued, routed and reported just like a normal call."

But, many companies don't want to handle calls these ways for one reason or another. The biggest reason seems to be that the video application is some company's first foray into a call center-like application, and they don't want to spend the big bucks for a sophisticated ACD that they will use for nothing else.

There are also a few applications (my friend the financial consultant comes to mind) where the company doesn't think of their video users as agents, doesn't have a lot of them in one place or has another good reason for wanting to keep the video "agents" away from the rest of the call center.

If you have just a handful of kiosks, you can handle the incoming video calls with simple rollover techniques. For example, the customer at the kiosk might get a busy message and be transferred to an agent for a voice-only session.

When you have just a few kiosks or other video calls coming in, these simple methods can work. When you start talking about hundreds of kiosks, a multimedia ACD of some type is needed. There are several multimedia ACDs being developed that are designed specifically for small, multimedia call centers.

With video, you not only have to worry about the technology in your call center, but also how your "caller" is going to contact you. Some people believe that PCs equipped with video cameras are going to be as standard as CD players and hi-fi speakers in the next few years, but your application may not be able to wait for that day.

It isn't practical to expect everyone who wants to communicate with your call center by video-to have all the equipment they need at home. A video kiosk provides that equipment in a public area, such as a mall or building lobby, or in a semi-public area such as in a bank branch or even a super-market.

A video kiosk can be as simple as a phone, a screen and a video camera, or it can be as complicated as an automated bank branch within clear plastic walls that includes a computer, an ATM card embosser, check printer, special paper stocks, scanner, several video cameras, card readers and comfy chairs.

Believe it or not, before you start considering equipment for either end of the video transmission you have to give a little thought to what will carry the signal in between. The basic choice is between ISDN and the Internet. The most important question to ask yourself when diving into a video implementation is if the Internet will give you the bandwidth that you need.

If you just wanted to do talking heads, either could work. The standard that uses the Internet (H.323) may not be as high quality as the ISDN standard (H.320), but when you factor in all the money you'll save by not ringing up ISDN charges each month, the Internet standard starts looking a lot better.

Those other applications (forms, spreadsheets) also eat up bandwidth, and depending on the application, having the customer sitting there while the application loads just doesn't cut it.

If you don't like these choices, wait a year. There will probably be better choices. The trick is to not wait too long. Video today is much like computer-telephone integration was eight or nine years ago. While there are some advantages to sitting back and seeing how things work out with video, there are also advantages to being the first in your industry to offer a really helpful service. Video is worth a second look.

Part Four

Making Sense of the Call

You don't just interact with customers; you have a chain of contact that includes suppliers on one end, customers on the other, partners in the middle.

Chapter Fifteen

More on Computer Telephony

A few chapters ago I talked about computer telephony, mostly from the point of view of its history, how it got where it is now, and what kinds of things it allows a call center to do.

When I think about a typical call entering a call center, most of the time I envision that call as an old-style switched voice call, a person trying to reach a person. But more and more, that voice call is only part of the picture. Nowadays, the call center is required to deal with voice interactions (still the primary mode of communication), but also a diverse range of alternate media: fax, digit input (IVR), speech rec input, email, Web surfing, even voice-over-IP (straight from the Internet).

The reason that the call center doesn't collapse under the pressure of trying to make sense of all these inputs is because of the solid framework that's been built up over the last ten years through CTI, or computer telephony integration. Starting with the coordination of simple voice and data, applications have been created that mix the information lurking behind the scenes in corporate databases with the split second call processing and switching that brings the voice call into the center (and to the agent's desk).

This voice/data linkage is what allows a call center to graduate from something as rudimentary as screen pop to the more complex activities, like combining email processing with call processing using the same agent group. Or connecting a Web visitor with a call center rep for a voice call. It's all possible because once you start to combine the huge resources of the back-end with the customer contact infrastructure on the front-end (i.e., the call center, plus the website), then you're basically forced to examine every input that comes in, parse it, and figure out what you know and don't know about the customer originating it. And then you are forced, in turn, to come up with some kind of organized system for prioritizing, measuring and analyzing all these different kinds of interactions. It gets frightfully complex in practice. But it's important to see the developing complexity that results: to remember that if you have a call center and you decide you want to start pushing emails through that center, there's got to be a way to make sure that emails don't sit in someone's box for a month while they are handling phone calls, and that all 40 people who are handling emails handle them according to the same set of business rules laid down company-wide.

To solve these problems, a whole new category of software has emerged in the last few years, sort of a hybrid between help desk, customer support, sales force automation, and enterprise resource planning — it's called customer relationship management software, or CRM. (You might also see it referred to as customer information systems, or CIS.)

For now, the point is simply that it's computer telephony, by now a hoary old term that's almost laughably imprecise, that makes it possible to connect the front-end of the interaction with the data resources on the back-end. And that's how you are able not just to take a call, but to *make sense of that call.*

Computer telephony actually makes itself felt throughout the enterprise, sometimes in unlikely places.

Naturally, it starts in the business's phone system. The core of a business communication system is usually a PBX (private branch exchange). It's the switch that connects to the phone lines and the telecom service providers on one end, and the internal phone sets (through extensions).

Until the early 1990s, PBXs were closed and highly proprietary. The only standards they adhered to were telco standards for routing calls around the phone network. Otherwise, if there was anything special you wanted your phone system to do, or connect to, you had to go through the vendor.

Occasionally, a vendor would authorize a third-party developer to create an application for some special purpose, like reporting or call accounting. The problem, though, is that every vendor who wanted to create one of these specialized apps had to create a different version for every major PBX line. That made them expensive, and it made upgrading a constant headache.

CTI was initially an effort to open up those PBXs, making them adhere to certain standards for internal call routing (notably those promulgated by Microsoft and, at that time, Novell). The theory was that if all a PBX had to do was pass calls back and forth between the phone network, it was really not much more than a glorified single-purpose computer anyway. And a call could be seen as another form of data being routed around a network. Vendors spent the rest of the decade trying to make telephone switches function more like servers on data networks.

According to Dialogic (which makes component boards used in many of these apps, and is owned by Intel), computer telephony applications need two types of resources: media processing and call control. Media processing is the transmission and reception of call content. For example, a voice response system is primarily media processing, since it plays voice prompts and listens for touch tone or speech responses.

By contrast, call control is the actual creation, tearing down, and routing of calls, and the monitoring of where calls are going. For example, a call center application typically routes calls to the correct service representative and then displays data screens for the representative when the call arrives.

At its heart, computer telephony is a way of making sure that when a call comes in, you can take the data that rides along with the call and look up something in a database based on where that call is coming from. So as soon as that call arrives, you can know who it is. You can know if that person is a customer, or not. What kind of a customer he is. Whether there are any open issues with that person, and so on. No matter if the person who answers the call is a customer service rep, a receptionist, a salesperson, or the owner of the company — they all have access to the same bucket of information.

This allows you to know your customers — what they like, what they don't like, what they've bought in the past, how they've been served, whether they like to talk to you by Web or by phone, and most important, how valuable they are to your organization.

This intertwining of networks allows you to overlay applications like the new CRM systems. So you can coordinate between the front-end of the customer interaction, and all the data about customers that resides in separate islands throughout a company. So you can perform complex analysis about customers, even to the extent of prioritizing them in real time, as they call — so that you can allocate just the right kinds of resources to helping them, and no more.

In all, you can act more flexibly, with more relevant information available when you need it. Increasingly, computer telephony technology is an enabling technology, letting software companies create applications that in turn let the companies that use them be more nimble, competitive and responsive to customers.

For his part, this gives the customer the freedom to contact you in any way that makes sense to him, at any hour, for any reason. *And doing so in a way that comes at little or no cost to you.* It's not that CTI doesn't cost money — in fact, it can be frightfully expensive, depending on what you hope to achieve. But it allows the cost of any given interaction to drop. Calls are shorter and more fruitful. More interactions through "free" (voice response, Web and email, particularly) modes mean that when a person does make a phone call, they are better prepared, armed with more information about their account or your products.

What makes this possible? Mature, stable computer telephony and data networking systems that work with a minimum of maintenance and installation cost, and that remain customer friendly.

Computer telephony products on the market increasingly look to the Internet as a way of transporting information, particularly as an interface at the desktop. Call center applications use Web browsers as the tool to retrieve back-end information about customers and pop it to the agent's desktop.

It's also a good tool to use for collaborative customer interactions, where the customer and someone at the company are both looking at the same screen-full of information. In these apps, the rep can guide the customer to particular Web pages, can track what pages the caller has seen and can even keep an audit trail of both sides of the screen-based call.

The dramatic rise of the Internet has spawned a series of sister technologies to computer telephony called *Internet Telephony*, which are concerned primarily with transporting telecom traffic over IP-based networks, and incorporating more IP technology into existing carrier networks.

This begs the question of whether we really need computer telephony in the first place. After all, if the Internet (and its derived intranets) are really going to replace more traditional LAN-based data networking, computer telephony may be seen as a transitional technology, to be superseded itself by something from the data networking side in short order.

Though this may be the long-term future, the fact is that for most pur-

QUICK TIPS

1. Key among the things you should find out about your prospective vendor is the availability — and the cost — of after-sales support. Installing CTI is one of the most complicated things you can do to a call center; you're certainly going to need help with it. And you're going to need to know how much you're spending on that support before you jump in.

2. Get your prospective vendor to ROI your specific purchase. That is, get them to sit there with a calculation matrix, plug in the variables, and show you exactly how much money it's going to save, and when. This can be critically important when dealing with upper management (when you are asked to explain why CTI costs so much) and with small systems integrators, who often promise better results than they deliver. Having it spelled out on a piece of paper can make all the difference.

3. Not all CTI is created equal. There are (still) competing standards, and a great deal of proprietary hardware and software out there. Make sure you know what standards are in use in the systems you choose, and that those standards mirror your priorities going forward over the next several years.

4. Questions to mull: How difficult is it going to be to get applications for the core platform? Do they have to be customized? Or are there pre-written off-the-shelf apps available from the vendor or their partners? The answers could save you money.

poses, the traditional CT-enabled PBX is going to be here for a long time, as the standard tool in the business telecom toolbox.

The Internet, where it plays a role, will be on the data side, coordinating the flow of information between the front- and back-ends of the transaction, working in tandem with the data network-aware switches.

Bob Fritzinger, vice president of Voice Technologies Group, says that small and medium sized companies thinking about buying CTI-enabled equipment face an interesting choice. "The requirement to have ultra high reliability phone systems hasn't changed," he says. "So do you buy conventional networking equipment and wait while the solutions mature? And while the companies that make them learn from mistakes?"

Fritzinger says that general purpose IP telephony will mature in two phases. First, it will make its way into the carrier networks. Then, the second phase will take it to small business. "You see an increasing consolidation of all these services into one box. Why buy separate machines, and deploy your own LAN, when you can two years from now, go buy a box and you have a business? If I was starting, I would focus investment on the data network. Treat the telephone net as ultimately disposable."

The inherent problem that CTI aims to address is the dissimilarity of the two networks. "Convergence allows for the promise of a smart network with a smart terminal," Fritzinger says. The lesson: don't be a guinea pig.

So, What to Do?

Oddly, it used to be a lot easier to talk about computer-telephony integration. That's because CTI was once a clearly defined set of technological tools, with specific, if abstruse, goals. CTI was just a way of connecting disparate systems and making sense of the results.

But the technology has changed and so too have our expectations of it, and where it will lead us.

Over time, CTI-borne apps became more important than the core components on which the technology premise was built. This is only natural, when you pause to think about it. As a "developer" technology, it's interesting mainly to developers. But as an application platform, it is of critical importance to every business.

Purchasers of CTI tools now have implementation choices impossible just a few years ago. The old (and well-deserved) knock on CTI — that it was over-hyped, took too long to implement, cost too much, required an army of consultants and systems integrators, and was invariably much more "custom" than you would have liked — has given way to a better model for designing and delivering applications. Specifically, the speed with which vendors are adopting the ASP (application service provider) model has put a lot of tools and features into the hands of small and medium sized businesses that could not afford the capital expenditure of the old "custom" model.

Call center tools have moved far beyond telephony, even beyond data networking, onto the desktop both inside and out of the call center itself.

What's more, these very fruitful technologies are being used to literally extend your company outward, and inward. You don't interact with just customers; you have a chain of contact that includes suppliers on one end, customers on the other, partners in the middle. What's sprung from the marriage of phone and data networks is something no one really expected — application suites that take the same basic pool of information (your data) and present it in a meaningful and secure way for anyone you think needs it.

Your enterprise can literally be extended in ways that were not possible before. Ten years ago, people thought of EDI — electronic data interchange — as a great tool for companies to manage the flow of information and transaction processing between them and their suppliers.

The only problem with EDI was that it required dedicated networking, software, and proprietary standards between all users. And you had to be as large as GM to command your suppliers to get on board.

Now, all that's changed. Any company or individual that you deal with can enter your extranet, see the information you make available in a form that's most meaningful to them. You can transact business and literally, be there for the people you have relationships with≤, no matter where in the supply chain they are.

For example, say a customer calls to place an order. She dials in and the call is intercepted by an automated response system, known in the trade as an IVR, for interactive voice response. That interception is the single most important act in the course of the call — because it's the company's best chance to get the customer to provide some kind of identifying information about herself and her needs. The IVR asks her several prioritizing questions: Do you need sales? Or service? Are you calling to check the status of a current order? And so on.

Once the direction of her needs is established, the system can ask for an identification code: an account number, order number, coupon code — literally any ID, so long as you have a database matching customers to those records lurking behind the scenes. Most companies do have these data repositories. The trouble, of course, is that the information is too often locked away where it's hard to access at the precise moment it would be most useful: during the call, in the hands of the rep.

It's useful because the more the company representative knows about the customer and her history:

· the more likely he is to satisfy her needs in the shortest possible time;

· the less likely he is to repeat a mistake in her handling;

· the easier it will be to use the extra time and the reservoir of good will to offer a cross-sell or up-sell opportunity without seeming completely crass, wooden and out of place.

Remember: any given call may turn out to be your last chance with a customer. If she decides never to call you again, you'll never know about it. She'll just go somewhere else and you'll never have any knowledge of what went wrong. The better equipped the rep is with information about what that customer likes and doesn't like, the more likely you'll be in the position to know when something is going wrong and you're on the verge of losing that customer.

The original problem was that the phone system and the computer network were parallel infrastructures that had no pathway of communication between them. And while LANs and networking topology grew up in the 1980s in an environment that stressed — and rewarded — openness and third-party development, the telecom hardware business did not. No one went out of their way to manufacture a business phone system that could talk to anyone else's. At that time, there was no business reason to do so, and there was no clamor from the user community to make it happen, either. In the largest installations, users could afford to make custom connections between their data and voice networks, but the need wasn't there because as we have seen in so many high-tech industries, the really useful applications always follow the development of the platform, and not the other way around.

Of course, there wouldn't be so many ways for customers to come calling if it weren't for the power of the Internet. It is common practice to stand and gape at the phenomenal transformative effect the Internet has had on our society. But in this one case, customer contact, there's no doubt that the effect has been real, and dramatic. Ten years ago customers had only one effective method of contacting a company. Whether they wanted to buy, ask for a catalog, complain, or check on something they'd already ordered, they had to pick up the phone and call.

And although the ability to reach people in their homes through phone contact turned many a local or regional business into national companies, it created its own limitations. The amount of information that can be delivered by phone is restricted by the fact that it's designed to handle voice only.

When married to the Internet, however, customer contact becomes more dynamic. People on either end of the transaction have access to more varied types of information at different stages of the contact. And the customer has far more options when it comes to making contact with a company. The choice becomes far more nuanced, and the company's headaches (and opportunities) multiply. These are some of the choices:

· Email, the Holy Grail of customer interactions. It's fast, free, detailed, and extremely open to all sorts of automated processing. Customer emails can be analyzed for key words, parsed and then routed to an appropriate person for handling. They can be tracked and managed in much the same way as phone calls. The key differences, though, being that they are not immediate, and that they come in through a separate, parallel network.

· "Call-me" or "click-to-call" buttons. Not so common as people thought they would be about five years ago, this contact channel involves creating a bridge between the website and the call center rep. The customer, finding something that sparks interest either in getting more information or making a purchase, clicks on a graphical button that prompts for a phone number (or spawns an applet) that in turn creates a phone connection. Once established, the rep and the customer can share a view of the website.

· More common now, and in many ways a better tool than the call-me buttons, is live-text chat. Chat has become pervasive on the Internet, through instant messaging clients (billions of instant messages are sent each day). This is a way for a company rep to communicate in real-time with a customer by typing text back and forth at each other. It sounds cumbersome, except that the tools that facilitate this include sophisticated scripting and "call" management features. The rep can view the screens that the customer sees, and can carry on several simultaneous chat sessions. Data on how productive a person can be doing this have not been collected, but the category is catching on like wildfire because it's a cheap and reliable alternative to phone calls, and it's got a higher service level than email (where the expected response time can vary from "instant" to "never").

In addition to these pure "contact" channels, you find the Internet manifesting itself in interesting hybrid tools like Internet-based PBXs. Netergy, for example, recently announced a tool called the *Advanced Telephony System*, which is an all-IP "iPBX" hosting system. That means that it allows companies to have their PBX, only the PBX is actually a virtual system residing off-premise, using the Internet as the network through which data and telephony travel. In some ways, this is the moral equivalent of Centrex from an earlier era, in that it brings back the days when the interesting tech happened off-site. In this case, however, the advantage is that it brings a much wider variety of possible technologies into play at a lower cost for a broader range of companies.

Why are these CTI, or CRM applications? Because in any of these situations, the customer record exists (or must be created) and that means reaching back into the database to coordinate between the information gathered by phone and that gathered by other means. The phone channel, after all, is never going to go away. If someone has sent an email, it's likely they're going to call as well. This lesson was learned years ago by the first group of companies to try to meld the Internet with customer service tools, the vendors of help desk software. It was discovered that people with customer service troubles tended to spray a company with as many different interactions as possible. If they made a call and fifteen minutes later they hadn't heard back, they'd call again. They'd send an email. Each call or email would result in its own trouble ticket and case being created. And in the reports, it would look like you had three times as many service issues as you really did. So procedures and integrations were created to ensure that the case was kept consistent no matter how many times — or through what channel — the customer came in.

The companies that had the most experience in this, the help desk vendors, were the ones that ultimately transformed themselves into CRM companies: Siebel, Clarify, Remedy, Vantive, and others. These companies started with service, and then went on to apply the lessons of Internet service delivery to other kinds of customer interactions.

Look at Ticketmaster as an example. They put in a front-end/back-end coordinated software system for their customer interactions. Ticketmaster's Web customers will be able to engage in live, real-time chats with reps to make specific inquiries about events, get up-to-date ticket availability and seating selections, and receive event information, which can be "pushed" directly from the reps' desktops to each customer's computer.

The software gives Ticketmaster the ability to route customer inquiries across any contact channel in Ticketmaster's multimedia contact center. If an inquiry is sent to a representative, gets the CSR with information on a customer's current and previous transaction history. Ticketmaster can then use that data to track patterns and derive the appropriate cross-selling or upselling campaign based on what the customer has done in the past. This is just one case of bringing the appropriate information to bear at the precise time; it doesn't matter where the information comes from; what matters is where it goes.

What this application does is guard against the isolation of useful bits of information in places where they can't be accessed. You never know what might be useful, or where useful things might originate (in the call center or the data network, with a customer or a supplier, in a financial analyst's models or in an order processing system), or who might be able to make use of them.

▪ The Ugly Truth About Customers

The trouble with technology is that often its buying, selling and implementation occurs in an informational vacuum. It might look great on an RFP, but sometimes companies are scared or baffled into a technology decision that might have little to do with how they could actually use those tools.

Customer contact is one of the most critical areas in which this "wing and a prayer" approach to technology management occurs. And the reason for that: companies have a wrong-headed notion of who is in charge of the interaction. It's not them, it's the customer.

Ask yourself why email is such a popular interaction mode now. Is it because companies decided that they were going to offer customers an alternative to the phone? Is it because companies wanted to save customers money? Does it have anything at all to do with actions taken by companies to please customers? No, no, and no. It has to do with the fact that customers visited websites and began sending emails. Often, they sent emails to companies as a direct result of poor response to phone inquiries. Customers think nothing of the double- or triple-interaction, making multiple inquiries for the same problem. And emails began piling up.

Businesses had to respond to the customer move to email, or risk losing customers. Just as with the phone ten years before, they had to begin building email-handling infrastructures in reaction. Once the cycle is in play for a few years, and some companies begin to prosper by excellently handling the new interaction mode, everyone follows suit and email tools are built into many existing phone and CRM products. But until those tools are widely available, you have a case where it's customers leading, and companies following the market. And it's the smarter companies that saw what was happening and responded to the customer's drive that came out ahead.

On the flip side, look at live-text chat. This is a fabulous and exciting technology. It's an amazing mix of cutting edge tech and a guess about what customers will want in the years to come. But instead of a customer-driven channel, it's a company-driven channel. Companies are betting that they can use chat tools to handle customers that come in over the Internet and do it at lower cost. But unlike email, customers are not pouring in droves. There are some who think this is a transitional state between the call center of today and the Internet-video-phone of a decade hence. And in the interim, even though the technology is strong and robust, businesses are not throwing huge sums of money at it, in quite the same way that they are to CRM and email handling tools. It's not customer-driven.

The end-user companies that have to install the technology that's been spawned from the CTI revolution are often dazzled by the technology, but afraid of their customers. If you make the wrong bet on what interaction channels your customers will use, you'll have sunk enormous sums into a losing technology. Buy anything ISDN lately?

Take a catalog retailer as an example. A customer looking at a catalog, thinking about buying, might want to look at the website while at the office, seeing if the prices online match the prices in the book. She might want to see if the merchandise offerings are expanded on line, or more limited.

Or, at home, she might want to call and talk to a person to place an order. She might want to ask specific questions about what's available, and when she can receive it. Having placed an order, she might want to check its status, but not wait on hold to talk to a person. She might call an automated line, enter in the order or customer number, and hear the shipping info. Or, back to the website for the same question, depending on where she is and how she feels at that moment.

Different locations, different tools, different modes of contact, all at the customer's discretion. From a cost point of view, the only one that actually costs much is the one where she touches a live rep, and that one was a revenue-generator.

And of course, if she got a catalog with a customer code on it, it would have been possible for each of those interactions to capture the code and keep a running record of how *all* those interactions proceeded, intertwined. That's the role of the CRM system.

But just as the customer feels the need for control of the interaction, so does the company feel like they own the information that comes from it, and so must control the flow of the interaction. Knowing where the customer likes to go provides a roadmap for where and when to offer that customer the opportunity to either buy something, or part with some significant piece of information that might lead to a sale later.

One tentative solution is "managed self-service" — like the literature web-site — where the customer can download a document or brochure, after giving over some contact information. This makes for a better customer, one prepared to buy more at a lower cost of acquisition.

In this dynamic tension, of course, both sides are right. It's a relationship, after all. The trick is to manage the flow so that the company maintains control, but the customer has the sense (a real sense, not a mirage) that she was making it happen.

When you look at the bewildering array of tools on the market for managing customer contact and coordinating voice and data, it's possible to

QUICK TIPS

1. Don't let customers use you as a crutch. Make sure you provide your customer base with adequate training, or a voice response unit as a front-end to answer routine questions. Don't be afraid to tell your customers that they are eligible for a certain kind of service and no more — unless they pay for it. It's marketing's job to make that look like a benefit.

2. Ask the right questions of your vendors. Make sure the features you request are features you actually need (or might need in the future) — and that those features really work the way you want them to. It's not enough just to ask a vendor if his product is "Web-enabled," for example. Many will say yes and leave it at that, not explaining that there are many ways to Web-enable a help desk. Purchasers should be careful not to simply go down a checklist of "wants" without knowing how they want the vendors to approach those wants.

3. Have you standardized on an enterprise-wide hardware platform, and if so, does that keep you from getting the software you need? Will it keep some people from having access to the data collected by your help desk? This is a good time to evaluate what you've already got (and might want to change).

4. Groupware features can be especially helpful in customer service software. Especially helpful are features that let users communicate: messaging, smart routing and forwarding, and automatic priority escalation, to name a few.

5. Don't "dead end" yourself with closed architectures. Look for a product that contains DDE (Dynamic Data Exchange), OLE (Object Linking and Embedding) and ODBC (Open Database Connectivity) standards, among others.

hold these two perfectly contradictory notions in your head with equal fervor, and be completely right about both.

Underneath the buzzwords and marketing hype lies a real, critical problem: customers are fickle and difficult, and it is surprisingly difficult and expensive to bring useful data to bear on the interaction at the point of contact. Note that word: useful data. For information to be meaningful, it needs to be parsed, analyzed and made available to the person who needs it. The data, and the person who needs it, are both moving targets.

Which highlights the another essential truth about all of this technology, and indeed the entire art and science of managing customer contacts. That truth is that technology is not a panacea. It's not the answer to everyone's problem. All the technology in the world isn't going to make anyone smarter, nicer, friendlier or less afraid of their supervisors. Throwing dollars at technology can be a shortsighted way to solve problems. Dollars invested in state-of-the-art CTI technology in 1998 would have been written off long ago. Dollars invested in state-of-the-art training for reps, business analysis or customer satisfaction surveys that same year would still be paying dividends today.

The human element — on the inside and on the customer side — is the most often overlooked. And it's important to remember that the people who come closest to your customers are often poorly trained and underpaid, making them the weak link that can overshadow and render useless millions of dollars in technology.

Skills-Based Routing

If there's one thing that distinguishes the call center today from the typical center five or ten years ago, it's that we now have so many more tools at our disposal to assess the nature and properties of the call as it's coming in.

Instead of having to be content with the raw data that can be supplied through ANI or DNIS, we have a whole tableaux of data-gathering options at the ready, most notably the customer-entered digits from any IVR system.

This is important in many respects, because the more you know about a call (and the caller) the better able you are to respond with the most appropriate resources. Which in turn lets you handle the call faster, better, and more productively.

And the best way to do that is to route a call to the agent most capable of handling that specific call — routing based on an agent's skill or combination of skills, like language, training, experience, or any mix of those and other factors.

Routing agents by skill delivers profound benefits to your center — but the technique is not without problems. Here's what you need to know about skills-based routing.

In theory, skills-based routing is a call center manager's dream come true: always handle the call based on exactly what that call demands. The agent who deals with the customer is precisely the right person for the job. It allows you to provide the ultimate in quality customer interactions.

But there is a flip side. With high quality comes a high price. Because when you classify agents by skills it can become a nightmare to determine exactly how many people you need at any given time. The traditional methods of workforce planning break down in the face of too many variables assigned to each call and each agent.

Skills-based routing (SBR) is one of the most requested features in new ACD purchases. That's not surprising — it gives the center manager much more flexibility in assigning agents to groups. It gives you precise tools to determine that if a call meets certain criteria, it goes to a particular group or person.

For example, you may set it up so that calls are separated by language. With an IVR front-end, the ACD can send Spanish-speaking callers to a set of agents who are fluent in that language. Or, you might send calls from priority customers to the most senior agents. Or calls for a particular service to agents trained in that area. There are few limits — which is what makes it very appealing. It gives the call center manager the ability to construct call flow patterns that — more than ever before — match the real needs of actual callers under their real business conditions.

Skills-based routing is a relatively new mechanism for matching a caller's needs with an agent who is capable of meeting those needs. In traditional call centers, agents are grouped into categories: each agent, regardless of his or her individual knowledge and capabilities, is organized into an ACD group which is assumed to be comprised of individuals with an equal level of knowledge and experience. When a call is answered by that group, theoretically, any agent is assumed to be equally equipped to handle that caller's needs.

In the real world, agents have very different capabilities or skills. For instance, a new employee may only have the experience to handle very basic types of calls. An experienced employee, on the other hand, may be able to handle a wide variety of call types. SBR essentially embraces this reality and views each agent as being unique.

■ It Gets Tricky

One of the obvious downsides is that it is very difficult to manage. There are simply too many variables that influence the quality of service experienced by a particular kind of caller and the degree of utilization experienced by different types of employees.

A fundamental characteristic of SBR is its ability to define and inventory the unique strengths of each individual. As callers request specific types of assistance, as defined by a variety of means, including VRU selections or DNIS digits dialed, SBR matches those needs with an individual capable of responding to them. In effect, SBR breaks down the tradition of assigning agents into ACD groups and views each agent as its own 'group'.

At its heart, skills-based routing is a system for distributing calls that come into an ACD. Traditional routing is based on two factors — an equitable distribution of calls among available agents, and the random nature of incoming calls. Skills-based routing changes this somewhat: it routes calls to the agent "best qualified" to handle the call, measuring "qualified" by agent parameters you set. The ACD does this in two steps. First, some front-end technology must be used to identify the needs of the caller. ANI, DNIS, IVR, speech recognition, even Web-based pre-prep. Then that information is matched against the sets of agent skill groups.

There are two ACD advances that let you run skills-based routing effectively:

· Leaving a call in an initial queue while simultaneously and continuously checking other agent groups for agent availability;

· Or allowing an agent to be logged on to more than one agent group (in this case a skill group) at a time, assigning priorities to those groups by skill type.

What is the impact on call center productivity using this style of routing? It can be significant, especially in call centers where there are many different types of calls being handled and many different mixes of employee skills. The impact is likely to be relatively small in a large call center that handles only a few different types of calls and has relatively few cross-trained employees.

The Problem?

When you increase the number of options for routing, you naturally increase the complexity of the calculation involved in figuring out the optimum call center configuration.

When you want to figure out how many people you'll need on staff during a particular period of time, or how many trunks, you might turn to a workforce planner. This kind of software takes known data, adds your expectations (parameters) and calculates the state of the call center during the period you need to know about. And it helps you create the operational things you need to base on that, like work schedules.

The problem is that most of these software systems have (until very recently) relied on the Erlang formula for analysis. And Erlang assumes that calls coming into the queue are random and unknown. But again, skills-based routing is not based in random and unknown patterns. It's based precisely on *knowing* something about the call. Once you know something about the identity or needs of the caller, you are moving the ACD's routing away from Erlang. In other words, Erlang-based predictions won't work well in call centers that use skills-based routing.

Some software vendors are trying to move beyond the limitations of Erlang calculations, with mixed results. Pipkins' Merlang, for example, is an attempt to move past Erlang to something that more accurately represents call center traffic. Other workforce managers from Aspect, Blue Pumpkin and others have tried varying approaches, and there has been a lot of debate about the use of an alternate and intertwined technology called "simulation" to overcome some of Erlang's shortcomings.

Jim Oberhelman of Bard Technologies says that Erlang is not useful at all in predicting for skills-based routing, because to do skill-routing you need to know what's in the queue.

Also, he says that predictions based on Erlang overstate requirements by an average of 3% to 7%, depending on the type of application. In larger, high volume sales centers, for example (like cataloguers), Erlang predictions tend to come out very close to reality. But in help desks, with their lower volume and longer talk time, predictions can vary quite a bit.

▨ What To Do?

A simulator actually models the center's traffic based on your parameters. In the case of *callLab*, Bard's ACD simulator, it generates random calls, each with identifiable qualities. You can examine what-if scenarios that explore all the possible skill groups you'd want to create.

Simulation can help you understand the effects on service level and cost of your routing schemes. For example, I asked Bard to run a simulation showing the relative changes in the same call center under three different routing plans. The center is a hypothetical one offering sales and service on four separate products. Also, callers can get any of those sales or service options in either French or English. They posited 100 seats with a varied arrangement of skills.

That works out to 16 potential skill categories (four products times two options times two languages). A given agent can be proficient in any one or more categories.

In the first example, calls are routed traditionally — one call type to one agent group. Here, the average speed of answer was high, 109 seconds. All the other numbers were also high, but not out of the norm for your typical call center.

In the second example, calls are routed to the most qualified available agent using skills-based routing. The numbers are markedly better by all measures here.

As a control, they also simulated a theoretical maximum scenario: what would happen if all the agents are assumed to be perfectly cross-trained to take all of the calls and are organized into a single agent group. This is the highest cost, probably impossible example.

The difference between examples two and three are striking — because they are not that different. The simulation shows that skill-routing is far more efficient than traditional routing, and not so far off the impossible-to-achieve maximum.

(Bard's simulations project that to get skills-like results with traditional groups, you'd have to add 50% more people.)

And the savings are not just in people — the numbers for trunk minutes are even more startling, a key stat for an inbound 800 call center.

It's clear that skills-based routing brings major improvement to call centers. And it's possible to quantify that by using simulators. They are a good planning and staffing tool — but they are not schedulers. They can tell you how many people you'll need in given circumstances. But they can't work out break schedules, or figure out how to slot in vacation time.

Also, with the increasing popularity (and sophistication) of distributed ACDs that use skills-based routing to send calls through the public network to a linked center, managers need better tools to measure performance.

Customer Relationship Management

Customer Relationship Management.
In general, it's the art and science of making customers happy. In practice, it's a relatively new category of software that combines front-end customer interactions with all the customer data lurking behind the scenes throughout the company.

This is one of the most interesting and growing markets out there right now. It's a confluence of several older software niches (telemarketing software, data mining/warehousing, help desk/customer support, and others) that are made more powerful in combination. They are also more powerful when applied together at the point of customer contact, that is, the call center.

CRM is a way of tying the front-end of the customer interaction with the vast resources of data on that customer that exists in the back-end (and usually outside the call center). From a customer point of view, that means that the agent has information ready on who the caller is and what dealings they have had in the past. It results in shorter calls, and better calls that generate more revenue and are more refined and purposeful. It's only since CTI has become mature that this is really possible to accomplish on a large scale; until recently, this has been one of those apps available only to the largest call centers, usually as an extremely expensive, slow customization. All that is changing as the technology improves.

This has gone forward only because call centering has ceased to be an arcane specialty, welcoming professionals who come from data networking and IT as well as telecom people. What it takes to run a modern, top-notch call center requires more in the way of people skills and organizational savvy than the ability to program a switch. The goals of the modern call center are vastly different, and more customer-focused:

· Create many points of interaction between company and customer.

· Minimize the cost of those interactions.

· Identify those non-customers who might become customers.

· Give everyone outside the company the tools to gather information about the company.

· Gather as much information about those customers as possible.

· Provide that information to anyone in the company who can use it.

The key words here are **company** and **customer**. The only thing that matters is bringing together the people with the money and the people with the product. The call center is the best tool ever invented to do that. What is not interesting any more is the call. On a low level of operation, you certainly want to measure calls, because that's the metric of productivity in a center — how many calls did these agents answer in how short a period of time, equals how much it cost you to run the center.

But what everyone outside the call center is measuring is customers, and the results of the interactions. The call center's role is shifting — from a place where calls get answered to the place where information is exchanged.

Precious few call centers have made this transition yet. So many are still operating according to the old rules that it may seem premature to talk about this radical shift. But the fact that the technology is out there renders the point moot — this transition will happen, is already happening. If CTI is only used in 2% of call centers (a low estimate), and there are 100,000 call centers in North America, there are still an awful lot of state-of-the-art centers out there testing new ways of operating in this data-centric environment. And you know where those CTI-enabled centers are: hidden away in plain sight — you touch them every time you call Schwab and speak the name of the stock quote you want; every time you track a package with FedEx's website, whenever an airline calls you to tell you your flight will be delayed. At America's top companies, the ones that have the most riding on every customer interaction.

In the ideal world, every customer will be handled as if she is the company's only customer. Every interaction she has with the company will be enlightened by all the relevant information about her needs and desires. All the data she has ever expressed about preferences and background will be saved, and brought to bear on the interaction intelligently. It is not simply a matter of collecting data and storing it in a back office database somewhere, where it is reported on and forgotten. It must travel in two directions. When a company knows something about a customer, it must use that knowledge to build the relationship between the two parties. And it must do so relentlessly, so that the customer reaches the point where she is doing business with the company because she has so much invested in the company that *it's simply easier than switching to a competitor.*

And when you do that for every single customer interaction, across all customers, whether it comes in the form of a phone call, a Web visit, an outbound telemarketing call or an in-store visit—that is the mass customization of service.

There are really four ways of practicing customer service. The first, most basic method, is to treat every call just as a ringing phone—somebody must answer it at some point. When lots of phones ring, that means trouble.

Then there is the traditional method of call-centering-as-usual. Those phones are ringing? Get an ACD to route them, separate the agents into groups, and by God we can handle the volume. The goal of this method is cost-containment.

Then it gets interesting. Companies realize that the call center offers an unprecedented opportunity to gather information, and use that data in a rudimentary way to get new customers, or sell existing customers more product. This is the old CTI paradigm: "Oh, I see you've bought a green shirt in the past, Mr. Dawson. We have a new line of green shirts, would you like to try one on sale?" Crude, but effective. What differentiates this third stage of service operation from the cost-containment stage is the more optimistic view of the call center as a corporate asset, rather than a cost-center.

Taken to its logical extreme, using current technology, you reach a point where every interaction is completely customized from the ground up. Service is designed as a seamlessly integrated component of the corporate-wide customer retention strategy. It's reflected in the kind of data gathered and the information that makes its way to the agent's screen. This is starting to percolate out into the public consciousness: look at the success of Levi's jeans that are computer designed to fit women; or websites that let the customer choose how to view them (Yahoo, for example). Once the processes are designed, the cost per interaction is negligible. The revenue opportunities are enormous.

And yet, most companies today are stranded somewhere at the beginning, between containing costs and starting to think about ways to add intelligence to their calls. Until they stop thinking in terms of calls and start thinking in terms of customers, that's where they will stay.

That's where CRM systems come in — they negotiate the exchange of information between front and back. They enforce consistency of message and interaction across all channels of input. That means that a customer who contacts you via email with a problem gets the same answer to a question as the person who calls. And when a customer has *both* called you and emailed you, the rep who handles either interaction knows about the other.

And of paramount importance (but not always in the front of the call center manager's mind) is the ability to parse the resulting information and make marketing sense out of it. So you can aggregate the knowledge about your customers — what they like, how they reach you, whether they are satisfied with your company's methods, products, services.

Where Is CRM?

There's no uniform way of presenting CRM software. The vendors come from all over the product spectrum — everyone wants to, needs to, get into the act. So you find switch vendors buying or building this software. And help desk companies trying to shed that label by extending their software from pure service apps into the sales and data management area. And more recently, you find products built just for this niche.

What they tend to have in common is that they are generally suite-driven, modular, and built to integrate well into other parts of the call center and corporate structure. So you'll have a module that does sales force automation, for example, and another that handles incoming email processing, and yet another for coordinating with the CTI middleware layer.

A "closed loop" management system guides a rep through a complex series of steps automatically without the need for training or maneuvering through menus or commands. An automated workflow engine delivers the right set of instructions and prompts based on the specific customer request and the business processes the company wishes to apply to that particular consumer.

There are four basic reasons why vendors take this approach:

· It provides access to an expanded customer base by enabling a greater number of interactions through a wide variety of media.

· It manages, tracks and reports on customer relationships across all electronic channels.

· It delivers consistent, personalized service across e-contact/voice channels.

· And it ensures a low cost of ownership with simple administration and customization tools.

This is a fairly typical explanation of how front-end/back-end integration products are supposed to work, and to my mind signifies the rapid maturation of this category.

This is clearly the way of the future, with the call center defined as a core function of an increasingly de-telephoned customer contact infrastructure. Over time, these software offerings will extend themselves out into other realms, like order processing, and Web services (this is already happening).

▓ Strategies for CRM Success

Researchers at Gartner have outlined four B2B CRM scenarios (all equally plausible, they say) that could play out over the next decade. They range from a world in which broad CRM expansion exists to a worst-case scenario in which CRM is "on life support" — i.e., not as much of a market factor as it is today. Gartner analysts said enterprises must prepare to react quickly to the varying scenarios or face possible drastic business consequences.

The two key critical uncertainties that will dictate the specific CRM scenario are customer/partner disclosure of information and global economic conditions. By monitoring those key indicators for uncertainty, enterprises will be in a better position to adapt their CRM strategies before it is too late.

"The key to any CRM strategy is the ability to acquire and effectively use customer and partner relationship information. However, to acquire that information, customers and partners must be willing to disclose it," said Rob DeSisto, vice president and research director for Gartner. "Disclosure ranges from complete to restricted. In a restricted environment, partners have a high level of distrust and will do anything to prevent end customer data to flow back through the demand chain. Government agencies will play a more prominent roll in end customer privacy issues. For complete disclosure, we expect to see the opposite, such as limited government involvement, high-demand chain collaboration and customers believing that benefits outweigh privacy concerns."

Gartner analysts said the health of the global economy is the other key uncertainty that dramatically affects CRM strategic decisions. Economic conditions dictate the ability of enterprises to invest in new capital and technology, and end customers' buying behavior.

Gartner analysts made three recommendations for building strong CRM in the B2B industry:

1. Incorporate scenario-based thinking into B2B CRM planning — B2B enterprises should have four- to six-month, long-term planning meetings focused on examining the effect critical success factors will have on their CRM strategies.

2. Collaborate now; invest in collaborative training tools to improve channel partner selling effectiveness.

3. Consider CRM vendors and technologies that support a distributed collaborative framework. Distributors should focus initial investments in CRM on technologies that affect the entire buying-to-delivery process.

IDC divides CRM applications into three segments: sales automation software, marketing automation software, and customer support and call center software. In 1999, the sales automation segment was the largest, just edging out customer support and call center applications. However, in 2000, the customer support and call center segment became the largest part of the market.

"It's no exaggeration to say Internet technology is having a profound impact on marketing — perhaps more so than on other CRM segments. In this new e-environment, marketing is taking on a new and highly strategic importance and becoming much more closely aligned with sales and customer service," said Mary Wardley, director of IDC's Customer Relationship Management Applications research.

Overall, the CRM market is very fragmented, and competition is vigorous and dynamic. No supplier beyond the top five has more than 2% market share.

Managing CRM's Terror

CRM is more than a buzzword: it is an attempt by people inside and outside the call center industry to come to grips with two things.

1. An explosion in the number of ways customers can connect to a company (with or without a call center in the middle of that connection).

2. An acknowledgment that there is a need to manage the customer, separate from managing the "call" or the "case" or the "interaction" itself.

But in the rush to push technology on an industry reeling from a decade of fast-changing tools and too-hyped concepts, a lot of people have been left wondering what CRM actually is in real world terms. And where it fits into their actual call center.

Let's begin with the obvious. CRM consists, mainly, of applications that control the customer interaction, and manage the customer's business with you. These apps can be small and specific, like dedicated systems for email routing and live-text chat.

Some of these apps are larger, more "suite"-oriented. That is, they combine those specialty apps with more back-end functions that parse information on the customer coming out of the databases behind the scenes. These tools attempted to deal with the whole picture of the customer interaction, rather than just the call.

They tried to establish a hierarchy of rules for dealing with customers, based on the rules of your business, and then apply those rules logically across the various types of interactions, from voice to email to fax and even direct mail response.

There are three complications that arise from this intelligent point of view. One is that you have to go outside the call center, and include all sorts of other parts of the company, where different priorities exist and different ways of measuring success are in force.

Two is that in many cases that logic and those rules didn't exist, meaning that they had to be created from scratch. Or mediated between competing sets of rules created by different parts of an organization. Either of which can be an interesting and fruitful exercise, if you have the strength of character to undertake it.

Three is that you come to realize that CRM, as I first defined it, needs to touch and be touched by nearly every other existing call center technology. It's a set of applications, so it needs to talk to the switching infrastructure. It depends heavily on CTI middleware, and all the specialty apps for screen pop and connectivity that CTI brings along. It's *extremely* important to touch the front-end, with its IVR and VRU and speech recognition and so on. Even if you're running both inbound and outbound, or heaven help you, offloading some of your calls to an outsourcer and having to combine data from those streams. Say you're running multiple call centers yourself, and using workforce management software to schedule agents among them. You can see where I'm going. Quality assessment, monitoring, same thing.

Any time you collect data, report on it, and use that data to make some decision about what's going to happen with the customer, and by extension with an agent or a call, your "CRM" is going to come into play with your traditional technology.

You're going to have to deal the confusion, and the technology headaches, for quite some time. It might be helpful to look at this as the natural outgrowth of computer telephony. CT was a developer technology, and by that I mean it was born from the desire of third party companies to add applications onto closed switches. CT was the conduit between the computer and the phone system. Now that you take that conduit for granted, you're faced with a multitude of applications that make use of the pipeline. They used to function separately, things like sales force automation and contact management and help desk software and telemarketing. Now they're all part of the same basket of customer contact issues.

As computer telephony fades (and it should — never was there more hype about a more boring and transitional toolset than the 90's rave over CTI), this mania over the customer is taking its place. I think that there's a lot more value to worrying over the customer relationship. And instead of worrying about isolated pieces of technology, which was what CT was about, now you can worry about two more complex ideas:

1. Isolated bits of information, and how to make sense of different kinds of data about customers.

2. And isolated processes for dealing with customers that are out of alignment. (By this I mean, for example, do you promise them one thing when they come to you by email, and promise them something different when they call you on the phone?)

What CRM really boils down to is this: know where you want to end up, what the relationship is supposed to be in a perfect world, and then work backward to decide what applications are going to get you there. And then bring all the tech components (in and out of the center) into alignment with that goal.

Pretty simple, right?

▓ So What's Wrong With CRM Today?

A 2000 Frost & Sullivan end-user satisfaction survey (the most dangerous kind of perceptual survey, because end-users are so darned unruly), reveals a distinct lack of product maturity and cost-effectiveness, as well as general disgruntlement amongst end-users about the performance and capability of the CRM solutions available.

"As users grow increasingly reliant on the functionality and the data integration offered by CRM products, vendors are gearing their efforts towards fulfilling the criteria applied by the end-user. The market is expected to experience a shift to electronic CRM (eCRM) which aims to embrace the Web as an access channel," says Frost analyst Andy Tanner Smith.

"As vendors begin to discuss their solutions in terms of the next wave of eCRM, the market is open for any vendor of sophisticated, yet sensibly priced offerings. The study emphasizes that the current crop of vendors falls some way short of this," he says.

The market is typified by fierce competition between vendors that have differentiated their products and services along a number of lines. These include, but are not limited to, product functionality, integration with existing systems, ease of use, and the range of target markets.

Frost & Sullivan suggests that CRM vendors need to practice much more of what they preach. CRM applications do not always live up to customer expectations and users needs are not fully catered to.

The highlights:

· While users expressed relative satisfaction with the technology, the presales consultancy and post-sales service they receive from their vendor could be improved. One of the most crucial factors in this context is price.

· The consensus among respondents was that CRM products were overpriced. However, CRM is an explosive market and prices are anticipated to somewhat decline in the medium term.

Tanner Smith: "Expectations have been raised through the high price tags attached to leading CRM products. Such high prices are not uncommon in emerging markets because the competitive advantage offered by the product lies in its ownership. When mere ownership of a CRM solution fails to provide that competitive advantage sought by users, they will look for complete out-of-the-box systems; improved and wider scopes of functionality; price reductions in order to achieve more rapid returns on investments and better services from vendors. Indeed, users will be looking for all the elements of a competitive market that are lacking in the vendors of CRM at present."

· Products being touted by the vendors as complete solutions fail to provide the features that clients require. Many vendors offer base products rooted in sales force automation, marketing automation or customer service technology and add functionality through third party vendors.

The integration between these disparate technologies is of key concern to users and potential users. As such, the researchers conclude, the absence of mature CRM offerings clearly inhibits market growth.

Similarly, a study from Ovum laid down some challenges for the CRM vendor community.

As organizations move rapidly to on-line commerce, they rely on software to provide the excellent customer service, which is critical to attracting and retaining customers. This, Ovum declares, is eCRM software. It helps them move away from cutthroat price competition and into profitability. But the research shows that the cadre of well-known CRM vendors is not able to respond to this need.

The ultimate goal of eCRM, at least as far as Ovum sees it, is to personalize the experience for customers who are using self-service channels to ensure they experience a consistency and quality of interactions that they'd expect from the best traditional channels (i.e., voiced agent calls).

Customers who buy through the Internet are demanding an experience that matches or is better than the physical world. They need to gather information, ask questions and buy products with the same ease as making a purchase over the telephone or walking into a shop. To retain loyal customers, a website must do more than just make information available and execute transactions efficiently. It must analyze knowledge gathered about customers to make sure information and offers are personalized to their needs.

Choosing the Right CRM Tools

CRM can be a mountain of headaches for attentive call center managers trying to understand what it means and how it should be used. There are tools, and concepts, that are sweeping through the industry and laying waste to a lot of old assumptions about how the interaction between company and customer should be handled.

In light of that, it helps to know before you take the plunge what you should expect to find in evaluating CRM tools. In my view, it all comes back to one basic idea: CRM is supposed to make it easier for the agent to deal with the customer. Nothing less should honorably be called CRM; nothing more is really required to do a satisfactory job.

To refresh: CRM is the blanket term for software that coordinates between three things:

1. The back-end databases and applications that hold (and use) the data about customer histories, product information and company policies.

2. The telephony and ecommerce infrastructures on the front-end that provide input for getting at the right back-end data.

3. And the agent desktop applications, which you could think of simply as the interface into all of those elements in #1 and #2.

When all the interactions were telephone calls, the glue in the middle was called *middleware*, and it was mainly for telecom pros to worry about. The call center worked just fine from an operational point of view with whatever the tech people said was appropriate for your switch. End of story.

But when you start factoring in email and Web interaction, with all its many permutations and the lightning fast changes in technology that are part of it, you need some higher power coordinating not just the tech, but the actual information running through your organization. CRM is supposed to be that higher power, making sense of all the data and putting a single screen in front of agents. It's also supposed to make the information accessible outside the call center, but that's a tale for another day.

When you decide the time is right to evaluate CRM systems, the first thing you're going to notice is that there are too many of them. Many will be gone in a few years. But you can't wait for that shakeout to happen — you have to decide now. The next thing you'll notice is that a lot of the products that call themselves CRM tools only loosely fit that definition. You will waste valuable time winnowing the field to exclude from "CRM" things like

CRM readerboards and CRM monitoring systems and CRM dialers. These are tools that *work in a CRM environment,* but precious few are CRM tools in their own right.

Your basic framework for discovery of the perfect agent-oriented CRM tool will be a system that connects with four things. Anything less than these four really shouldn't count for most people's purposes. They are, in no particular order: the ACD (or other switch, like a PBX), the telephone network, the IP network (that is, the Internet), and the existing CTI middleware that's probably in place. Anything less is really more of a value-add to a CRM system or traditional call center setup than a true CRM platform.

You'll also find yourself juxtaposing that framework against the traditional spectrum of size and expense. At every step of the way, for example, you'll be thinking about whether it makes more sense to build your own customer system, buy a tool built just for call centers (perhaps from a vendor who provides one of the four elements just mentioned), go all out and convert to an enterprise-wide system, or even choose something delivered through an application service provider or other outsourcer.

Those are tough choices, motivated by a lot of factors including price, company culture, your view of the future and your sense of how hard it's going to be for your existing infrastructure to keep up with the tech needs of these new and growing interactions. Enterprise-wide is perhaps the riskiest, and costliest, way to go. But there's a strong argument to be made that it's the best for the long term, because it has the potential to create a company-wide culture built around the idea of better, more productive customer interactions. It leaves no one out, creates no islands in which bad behavior and unaccountability can persist.

ASP-delivered services are a good hedge, for those who think their company isn't ready for an enterprise-wide system, but know that they need to invest in the highest level features, perhaps for competitive reasons. There are almost no circumstances left in which it's better to roll your own application. (If there are, please tell me.) It seems to me that creating an in-house, custom tool gives you a chance to reinforce old, bad habits, and robs you of a chance to take advantage of the very real technical expertise and innovative ferment going on outside. Yes, you may have very good consultants and systems integrators do the job, but by

2005 or 2006 you'll be shopping again, and by then you'll be too far behind the technological curve to continue to build custom; you're just postponing the inevitable, as far as I can tell.

These, so far, are issues of corporate culture and philosophy. But just as important are hearing from — and appealing to — the various internal constituencies that are going to be impacted by your new CRM tools.

Management, for example, is going to need the system to do and have certain things. Key among the concerns in this quarter are going to be reporting, and ensuring that business rules are applied with consistency across the various types of interactions. That is, managers are going to want to be able to determine, somehow, that emails are costing x cents per interaction to handle and calls are costing y cents, and what is the satisfaction rate across both types, and what is the relationship between cost and result, and all that wonderful stuff.

Management will also be keenly interested in the way CRM tools affect quality control through agent performance measurement, especially through hooks to monitoring, training, coaching and assessment tools, and through further hooks leading to data mining tools. (These areas are increasingly related technologically.)

From the agent point of view, you'll be looking for ease of use, with screens and menus that are easily customized, and yet consistent from user to user, or within a department. Almost every tool worth looking at is going to have a browser layer for accessing information. This could show up on the agent's desktop, or on the view shown to managers, or both. You're also going to be interested in exploring the way the system handles scripting, and again, the way scripting is affected by multichannel interactions: the way it walks agents through phone calls that are really collaborative Web sessions, for example, or live text chat interactions.

Ease of use is incredibly important because of the need to reduce training costs, reduce turnover, and in a high turnover environment, to make it easier to get entry-level people on the system in the shortest time possible — all while ensuring service that's consistent and consistently good.

The third constituency that needs to be appeased (or catered to, depending on your point of view) is the IT infrastructure. These folks are involved in call center planning like never before. In the old days, their influence would stop at the end of the telephone wire — they were involved in how the telecom flowed in the center, and how the LAN was wired, but not much in the call flow dynamics and not at all in the business decisions about how to handle customers. Now, all these things are wrapped together thanks to voice/data convergence and the affect of ecommerce on customer actions.

Since everything is now deeply involved in a company's core data and networking infrastructures, IT is critically important to supporting a company-wide effort to keep the data flowing and keep it intact, and consistent. Everything from the database management to agent training, analysis and reporting flows through their bailiwick. Any CRM system you choose will be primarily under their control. In fact, as you read this, it is increasingly likely that the "you" I am writing to is an IT professional instead of an old-style call center pro.

When all is said and done, there are a lot of factors that go into the selection of a CRM system. Too many RFPs are half-assed, and driven by the fear of a future that might never arrive — insisting on features that will never be used as hedges against uncertainty.

Instead, pay attention to what you do, and what you need, and what you can reasonably expect your customers to want from you. CRM is a truly transformative technology, but it's not smoke and mirrors, and it's being offered up as both a miracle cure and a business necessity. It's not any of that. It's a set of premises, not promises. You need to do a deep analysis of what you want your customer contacts to be like, and how much you want them to cost, and what they will generate in revenue, and how far you're willing to go to get to those goals. These are the things that you should be doing anyway, because they will create a better call center environment, CRM or not.

Order Processing

Order processing systems give you power over your inventory and pricing. They help you sell more in any given phone call. They take the guesswork out of the order-taking process.

The work is not over once you've made the sale. In fact, it's just beginning. The productive call center has automated call routing and the customer data. Why stop there?

Processing orders is the bread and butter of many businesses. If you are going to spend good money to deliver good service, you should pay attention to how you deliver the goods as well.

Yes, order processing software can make your order department more efficient. Yes, it can help your reps take orders faster and more accurately. And when things go wrong, it can certainly cut the time required to search for records of past orders from minutes or hours to just seconds.

It's true that all the time you'll save using order processing software could save you money on your toll free service.

While those are all great reasons to buy order processing software, there is a much better one: The information you gather while taking an order can help you offer better customer service and increase your sales.

Software systems make a difference in how you track sales and manage inventory. They can collect data and show you how well your products are doing.

It boils down to control: the call center with control of its sales is only half finished. Here are some of the things you can do with order processing systems to control the order process.

With many order processing packages, it's easy to generate pick lists and mailing labels from the data entered on the order screen. This makes things easier for fulfillment and means your orders get shipped faster.

More accurate order-taking means fewer complaints to customer service. When there are problems, your automated records let you resolve them on the first call.

Automation means that when your order department learns the address and telephone numbers of new customers, or the new addresses and telephone numbers for old customers, the information can be used by your list people for more accurate direct mail or telemarketing efforts.

Information about which customers are buying which products, and which customers are due for a re-order, is valuable to your sales and marketing departments.

If you've automated the rest of your sales cycle but are neglecting order processing, the time has come to fill the gap. Here's what to think about when looking for an order processing system for your call center.

1. Collect data about customers. Like other types of software, much order processing software is built around databases that hold information about your callers. They join the call records to the product records. That makes it easier to see the big picture. It also helps you deal with calls. It's critical that in the move to implement high-powered back-end/front-end integration systems, the CRM products, you not forget to connect processes like order-taking and fulfillment.

If a customer is already on the line, an agent may need to review the history of that account. This kind of software does more than just take orders — it enhances customer service. So you can ask a customer 'how did you like that blouse you bought back in April?' The customer feels like she is getting personalized service, and she is.

Customer data is also important because it speeds each call. If you get a record out of the database and onto the agent's screen using just a name or account number, it eliminates repetitive re-keying of data every time that person calls. And maintaining a history of previous payment methods speeds up ordering.

2. Micro-manage your inventory, pricing and product development. Someone calls to place an order. You should be able to tell them whether that product is available, when it can ship and describe important features of it.

Good software tracks the products you sell most (and least). Turn that information over to marketing and development, so they can figure out what's working (or not working). For business-to-business users with regular customers, include details of special pricing plans. The software you buy should have the ability to calculate volume or preferential discounts.

When a caller has a question about a product, how will your reps find the information they need? Must they exit the order processing software to get to information in another software system?

It's important that your reps be able to retrieve some product information right from the order processing system so they can answer customer questions while taking an order.

As with product information, it's important that your agents are able to quickly check if ordered products are in inventory, without leaving the order processing program.

Depending on your company's inventory tracking methods, it may also be helpful if ordered products are deleted from your inventory records as soon as they are ordered.

Some order processing software packages have quite sophisticated inventory tracking functions, either as part of the basic package, or as optional modules. For example, it's helpful for the rep to know things like:

· if items are available

· when a back-ordered shipment should arrive

· if you are awaiting drop-shipment or if the item has been drop-shipped.

3. Use each call as an opportunity to cross-sell or up-sell. Some order processing systems do double duty with call-management features. Users can create scripts to increase sales with each order. It's one way to keep that customer from walking away without buying.

4. Keep track of your sales data and lists. Your order processing system must have good reporting and tracking facilities. The information you get out of this software is just as important as what comes out of your ACD. When purchasing a system, ask what reports it generates. These are a few important ones:

· How much product you sell by source (catalog, ad or special promotion).

· Product profit reports.

· Comparison of sales month to month, by product and by agent.

5. Integration with other sales-cycle software. The goal is not merely to automate the ordering process, but to create a cycle of information that can increase sales and improve customer service. The ability to share databases and other information between software packages is crucial.

If you already have sales or other software in place, it may be more effective to find an independent order processing package that can share information with your existing systems.

Order processing software now commonly includes features traditionally found in telemarketing software, including literature fulfillment and post-sale follow-up.

As data about your products is collected over time, augment it with information about the customers who buy them. Reorders are critical to staying in business. Software like this helps you develop controlled, targeted lists to keep the customers you have.

6. Transfer data to other in-house departments. The free-flow of information makes your organization more productive. When an agent takes an order, the entire process is sped up.

The details of the order are added to the customer record. Sales and marketing benefit. That order is electronically routed to the shipping department, where an instant assessment of stock takes place. Better management of inventory and shipping means money saved.

Software links to accounting and accounts receivable mean fewer errors. No one has to decipher or re-key details from a form when the data is transferred automatically.

Another benefit is forecasting. Information gets to financial departments as quickly as orders are placed. They can make more accurate — and more timely — projections based on daily sales.

If you use an outside fulfillment service, an order processing package can let you can send them your orders electronically.

7. Telephone integration. The ability to deliver screens of computer information along with a telephone call saves time — and time is money when you are paying the freight on toll free lines.

   Find out what kind of telephone integration is available with the order processing software of your choice. It should be easy to find sophisticated integration features on programs designed for larger computer systems like minis and mainframes.

8. It should keep track of all the major shipping systems. Whether you use UPS, FedEx or any other shipper, software gives you more flexibility.

   It should set up each carrier's rate and zone information, necessary forms and paperwork. And, it should be able to make any updates behind the scenes, using the Internet. In fact, as with any enterprise software these days, the closer the user interface is to a Web browser, the better off you will be in the long run.

   Use inventory tracking features to record the weight of each product. Combined with the shipper's tables, you can tell a caller the exact shipping weight, cost of delivery and estimated date of arrival.

9. Interface with credit card authorization systems. Taking credit card purchases over the phone requires you to verify the authorization of that

sale with the credit company.

Many call centers, with their dozens of simultaneous sales, use a third-party credit card authorization service. They send credit card orders in batch mode several times a day and verify them.

Your big decision is whether to go with instant credit card authorization, which is faster and helps weed out fraudulent orders, or batch processing which lets you use your agent's time and your telephone lines more efficiently.

10. All the other little things. Order processing systems contain all sorts of neat little features to make your agents' lives easier.

For example, make sure it assigns sales tax based on zip code or locale. This kind of thing is especially helpful if you have a corporate presence in many locations. One nice feature I've seen that's pretty standard now is the one that lets agents enter a zip code and have the city and state fields filled automatically. It can be a real time-saver.

Browse through some feature lists or stroll through some brochures. Features like these are not major selling points, so you can miss them in a targeted search. One of these "nice to have" features might become a "must" for your operation.

Order processing may never have the flash of sales or the warm-fuzzies of customer service. But with order processing software, it can be an important part of your sales cycle, helping both sales and customer service, and winning friends for your company on its own.

QUICK TIPS

1. One way to keep customers happy is to be upfront with the truth. You should immediately acknowledge back-orders or out-of-stock products. Be as specific as possible about the actual shipping date.
2. Never fulfill an order without going for an additional sale. Let customers know about other products or services that might interest them.

Part Five

Critical Peripherals

It's an easy matter to quantify the
time people spend on hold. The
whole notion of service level is
derived from the idea of answering
a certain number of calls within a
certain amount of time. Behind
that, of course, is the idea that the
people who are not being served
are sitting a virtual waiting room
until their number is called.

Readerboards and Display Technologies

Display boards, readerboards, are a stable, versatile technology, and an inexpensive way to quickly improve call center performance. They are also rapidly becoming if not obsolete, than at least not as necessary as they used to be, replaceable by more malleable, "soft" boards that sit on the agents' desktop screens.

Most of the talk you hear about managing your call center employees focuses on agents — do they have enough say, control or information. But do you ever think about the supervisors?

For quite a while, the case for readerboards has been that they put useful data out there on the call center floor for agents to see it, where they can use it to make better decisions about how to do their job.

That data also helps supervisors, who are no longer chained to the ACD stat terminal, who can get up and help people individually at their desks, without fear that an alarm threshold will pop up as soon as they walk away.

Supervisors spend less time playing catch up, more time doing people-work than babysitting the numbers. Agents feel less like they're being treated like children. Everyone likes their job better and turnover goes down.

End of story? No, not when you realize that the possibilities in reader-boards are broader than you think. There's a lot you can do with them, and a lot of ways to configure them. It's not just a matter of throwing one up on the wall and watching data scroll by.

▦ The Benefits

Today, display technology mostly comes in three flavors: the traditional red/yellow/green readerboard, hung on a wall for many agents to view at once; a TV/monitor version of the remote board; and a stat readout that's fed directly to each rep's desktop PC. For discussion purposes, I'll call them all "readerboards" unless I specifically mean a desktop software data scroll.

Whichever type of system you prefer, there are four major benefits you can count on when you implement some form of display technology:

1. Improved call statistics. Simply put, when people know what's going on, they respond faster. If agents know how many calls are in queue, or that an alarm threshold has been reached, they are less likely to go on break at that moment. They are likely to speed through post-call work, or put off optional assignments until the peak smoothes out. The readerboard then has a hand in bettering the stats, not just reporting them.

2. Enhanced supervisory time. Get the supervisor out from behind a desk, and onto the floor where she can answer questions. Readerboards, which usually work in tandem with an ACD stat terminal sitting on or near the supervisor's desk, give managers flexibility. They have time to monitor more calls, improving call quality. They can confer with agents about specific problems, or give better training and guidance.

This is possible because they know, as they move around the center, exactly what the ACD status is. With most models, they can program the display board to show rotating groups of statistics (for different ACD splits, for example). Alarms can trigger bright colors, audible tones or other special effects.

3. Improved morale and an empowered workforce. Agents know about problems *as they are happening* — not later, when they can't do anything about it. The more they know, the less likely they are to have serious complaints about their work, and that can keep them in the job longer.

Lowering turnover reduces training costs. It also improves service by establishing a core of experienced reps. Agents feel like they are in control of their job, their actions, and most important, that they control the numbers (not the other way around).

According to one survey of readerboard users, service reps improved their self-management by 17%. (And respondents were able to decrease staff levels 4% due to greater efficiencies.)

4. Readerboards allow you to achieve other goals. Say you want to allocate some agents to a part-time outbound program, for follow-up or post-sale calls, for example. Without some kind of display, the supervisor has to control when the agents flip between inbound and outbound.

But a call center display system gives agents the ability to know that when calls are low, they can move over to outbound and make a few calls. And when a surge of calls comes in, they see it and can flip back to handle the queue. It takes a lot of the pressure off both the agent and the supervisor.

Buying Smart

Here's some advice on what to look for when shopping for a readerboard or other display system:

· Make sure it has an audible alarm at a threshold you choose — what's the point of having it flash and blink in colors if nobody's looking in that direction?

· Three-color LEDs that change color in response to thresholds you set.

· A refresh rate (how fast the data is changed) that is fast enough for your needs. Some displays refresh data in 30 seconds. Some call centers have high abandon rates at 30 seconds. The two wouldn't work well together. This is also going to depend on the software that drives it. A readerboard is, after all, only as useful as the data that comes out of the switch and the software that formats it for viewing. Which leads to the next point:

· Make sure that the switch, the reporting toolkit, the readerboard and its driver software all come from compatible vendors. This is not as big a problem as it used to be, especially if you're looking at a "soft" PC-based display system.

· Displays that can be individually addressed and also work in a network of linked displays.

· The ability to schedule messages, to save messages and call them up later, and to have a message sent after a certain threshold is reached.

■ What's Out There

I'm not going into details here on the technology of the old-style reader-board, because that's a technology that's either readily understood, or unnecessary. Buy a board, plug it in, configure software, type in prefab messages and you're off and running. I'll wager that the hardest part is figuring out which wall to hang it on.

No, what's really interesting are the PC-based displays. In the past, people didn't want to use them, because they took up "valuable real estate" on the rep's desktop. That was in the days of the 14-inch monitor. And because "all that data will confuse the reps" — someone really said that to me once. (Bet the turnover in that center was pretty high.)

Since the underlying technology — the software — is fluid, it really doesn't make much difference what form the output takes. PCs, especially in a browser-based world, really do make the most sense.

For example, take Fujitsu's *IntelliAgent*, a desktop application that functions as a local readerboard for the agent's PC screen. It provides all the stuff you'd see on a traditional wallboard, including personal performance information and group statistics.

Because *IntelliAgent* is a desktop app, it includes call control features, like letting the agent view the status of other agents before transferring a call to them. It works with the *IntelliCenter* management package, and incorporates up to 50 definable templates which administrators can customize for different agents or groups. Information is presented through color-coded thresholds and icons.

Supervisor terminals have had this kind of information for years; it's interesting that this is migrating to the agent desktop at precisely the moment that call centers are confronting issues of agent empowerment and high turnover. Putting data directly into their hands helps them manage their own workflow and make intelligent decisions about how they do their jobs. That's got to have long term benefits.

Desktop messaging is increasingly a cost-effective alternative to physical readerboards. For one thing, they offer a degree of programmability and flexibility that hard-wired boards just can't manage. One such product that was originally developed for the call center is now being marketed as a company-wide broadcast messaging system for intranets.

Softbase's *NetLert* puts its messages into a user-sized scrolling marquee on the desktop. Its messages can be pulled by users directly on the system, or by remote users and telecommuters. It's good for group memos, and for communicating basic PBX data output to a wide number of users. The client application periodically polls the data source for refreshed information. Obviously, this is not the same as a system that pulls data every ten seconds out of an ACD, but for an informal or lightly distributed center, where phone reps are spread widely within a single building, for example, it can be an effective tool.

It has more application as a "messaging lite" tool for reps who can't stop what they're doing to read incoming email than it does as a traditional threshold alert tool ("No breaks-the queue is full!"). But the future of wall-boards is surely pointing to the desktop.

CallCenter Technology's *Prism* is a data management tool that lets agents, supervisors and managers gather and view real time and historical data from multiple data sources. Like a readerboard system, *Prism* gathers information from the ACD; this is displayed on the desktop, though, rather than on a physical wallboard.

Prism then is able to coordinate it with exception data from workforce management systems, performance scores from quality monitoring systems and actual transaction data. What you get when you put this all together is a more holistic view of what's going on in the center, the kind of thing people in management are likely to want to see. When you put things together that aren't often combined (like agent performance and transaction data), you often end up creating new metrics that are more meaningful than the traditional cookie-cutter call center stats.

Another interesting feature is that *Prism* incorporates CAD drawings of call center floor plans — so supervisors can locate in physical space the data events that might get lost in columns of data. The system consists of a server and Console desktop software. Reports and data views come preconfigured for several different user points of view, and are largely customizable.

So where does this sort of thing leave the traditional call center? There are still plenty of places to get the hearty old readerboard. Companies like Telecorp Products, Texas Digital, Pyderion, Symon, Centergistic — all of these are working with the software that drives boards (or the boards themselves) and in nearly all cases, the results are far more flexible, more data-centric, than they were before.

Headsets

Headsets are a key ingredient in every call cen-
ter. They provide hands-free operation for your reps, letting them type faster,
talk on the phone longer, and walk away without annoying neck pain.

They are also the most abused, wear-intensive items in your center.
People swing them around by the cord. They yank them out of the socket.
They throw them in the drawer. You'll buy them this year. And next year.
And the year after that.

Luckily, vendors keep upgrading them to provide more comfort and durability. Warrantees have gotten longer. Prices are coming down, too. Most
good headsets average between $120 and $250, depending on features and
the volume you purchase. By some estimates, the average life of a headset is
about 29 months. And they nearly all come with warrantees that act as a
hedge against failure during that period.

Consider: if your center seats hundreds of agents, you could be spending
thousands of dollars each year for new and replacement headsets, plus backups.

It's not true that all headsets are alike. They vary in style, sound quality, and
technology (to some extent). Here's what you should look for when buying.

Which style? There are two main categories: monaural (one ear) or binaural
(two ears). Monaural comes in over-the-head, over-the ear and in-the-ear styles.

A lot of this choice will depend on individual comfort. Some people don't like the feel of one or more of these styles. In-the-ear could feel like it's going to pop out, or you might feel lopsided with monaural.

If it's at all possible, try to give your staff a choice of styles. Not only does that help the reps feel like they have some control over their work environment, but it lets them choose based on personal criteria: "this one will mess up my hair and that one won't," or even "this one sounds better than that one."

Also, pay attention to headset complaints. If reps tell you something is wrong with the headsets you've already got, take them seriously. Physical discomfort or poor sound, if left uncorrected, will affect your customer service in the form of higher turnover and bad morale.

Consider noise levels. Base your headset selections on the real world conditions inside your center.

Is it very noisy? Then choose a binaural model with a noise-canceling microphone. But if your reps need to consult with one another during or between calls, then a monaural might be better. That way they don't have to take the headset off and put it on every time they need to speak to someone nearby (which itself can reduce it's lifespan).

Sound quality. The caller should never know that they're talking to someone on a headset. They should also never hear much of the ambient noise of the rest of the call center (other agents talking, or a keyboard clacking).

That's why noise-canceling microphones come in handy. They raise the rep's voice while reducing the background noise. This is becoming a standard feature on call center headsets; at a recent trade show, most vendors were showing headsets equipped with flexible, noise-canceling mics.

A good tip when selecting headsets: find out if you can buy (and get a good deal on) accessories (another way of saying "spare parts"). The best way to get value out of your headsets is to extend their life through repair. Buy extra foam ear pads, eartips and connector cords. Replacing these relatively cheap parts keeps you from buying a whole new headset.

Convertibility. Some headsets are modular — that is, they let you reconfigure the way they're worn, switching between headband style and contour style. (This is a way to get around the impracticality of letting everyone choose their own headset. One model can suit many needs.)

If you go this route, though, be careful to check the construction of the headset and make sure it's rugged enough in all its configurations.

Universal amplifiers. These are great because they make your headsets compatible with different phone systems.

Check carefully to make sure the amp you want is truly universal. Some vendors make amps specific to one phone system or another. Ideally, the headset should move from phone to phone with no trouble. Also, make sure you know how to adjust the amp yourself, without relying on a technical person to set dip-switches and controls.

Headsets are a mature technology, with little fundamental change from year to year. What changes are hidden features, like warranty length, durability, and sometimes design.

As long as you buy based on comfort and durability you won't be making a mistake.

▒ How To Make Them Last

· Clean the cushions and voice tubes regularly; these parts accumulate grime and dust.

· Start a program to educate your employees in good headset care. They should know what it costs to replace and repair headsets. They should be made responsible for their maintenance.

· Keep extras on hand — extra headsets and spare parts (if the vendor makes them available). They are cheap enough to justify the expense.

· If the headset uses eartips, use the proper size. Women should start with Size 3, men with Size 4, and the eartip must fit into the outer ear canal.

· If it uses earpads, make sure the cushion is centered over the ear, that hair is out of the way and that the foam is in good condition. Replace it as soon as it becomes torn.

· If headset transmission drops over time, try cleaning the microphone screen or voice tube. Sometimes they get clogged with debris.

· If you're in a noisy environment, you're best suited for a binaural unit with a noise-canceling feature. For small offices a monaural or over the ear headset will work well.

· Look for clarity and volume. A headset with mute switch and a quick disconnect are nice features; they allow the user to communicate with someone in the room without the caller hearing, and allow for some freedom for the user to get up from their workstation without removing the headset.

Think you've heard all you can hear about headsets? Well, I found one I really liked, that was comfortable, seems durable, and fits snugly without getting annoying after several hours on the phone.

GN Netcom's *Profile SureFit* was a surprise. It's got no headband (unless you want one, the pieces are all modular, and you can rearrange the fit to whatever shape you want). It's an on-the-ear model, with the earhook holding the speaker and microphone.

The earhook is what makes it special, a comfortable composite material that can be manipulated to fit right; you can bend it, and it won't break, but it will retain its shape.

Headsets are one of those items that people have to choose for themselves, but I'm pretty sure that within the next year or two, models from lots of manufacturers will be mimicking the bendable, non-plastic feel of the *SureFit*.

It's not the only set out there that I like, but it's emphasizes how personal a choice this should be.

QUICK TIPS

Here's something to look out for when you're switching over from handset to headset. Sometimes, the headset lets you control the volume in both the headset and the handset. One headset user reports that when the headset was connected, the volume was too low on the handset, and couldn't be adjusted. If you plan on using both, make sure the volume can be set properly through both instruments.

On-Hold Messaging

Call centers are good at managing the customer relationship, and they are good at the mechanics of call handling and transaction processing. But sometimes less attention is paid to the mystery period in between — that time that the customer spends on hold, waiting in the queue.

We know a lot about the hold queue from a technology standpoint. It's an easy matter to quantify the time people spend on hold. The whole notion of service level is derived from the idea of answering a certain number of calls within a certain amount of time.

Behind that, of course, is the idea that the people who are not being served are sitting in a virtual waiting room until their number is called.

What do we do with those people?

The options available include:

· playing them messages and/or music to enhance (or ameliorate) the experience;

· allowing them to leave the queue to perform some other action, like querying an IVR system for basic information; and

· doing nothing at all.

Not too many people are going to argue that you should do nothing, but in a lot of cases that's what happens. Particularly when the time spent in the queue is too short to allow for any meaningful activity. We should count this as a success, obviously.

Playing messages on hold is an old idea that is still popular, with good reason. One executive at a music-on-hold production company once told me that if you look at messages on hold as a form of advertising, it has the lowest cost per impression of any comparable media.

Under those circumstances, then, what you want to do with the hold queue is use it as a value opportunity — tell people about what you have to offer, bring them information about your company that they might not otherwise be exposed to.

The hold queue exists because of an imbalance between the volume of customers trying to reach you and the resources allocated for handling those customers. If you have an equivalence between those two factors, then the hold queue disappears, but your costs go up too high.

Therefore, you want to build in a certain tolerance level. You need to assess, realistically, what kind of waiting time your customers will be willing to absorb, and under what circumstances.

If they want to buy something from you, then you want to answer their call immediately. The more valuable the customer, or the potential transaction, the lower the tolerance the customer will have for waiting in queue. So if you are selling $5,000 cruises to people who have booked in the past, answer immediately.

Likewise, you want to take the call from the Platinum card holder, and the Super frequent flyer, and the guy who has been to your website and pulled down all the information he needs to decide between buying your widget and your competitors widget.

In the same spirit, you probably also want to minimize the hold times of people who owe you money. The collections environment has a call center dynamic all its own. But it's pretty clear that when someone does respond to a collections mailing by calling in, the worst possible thing you can do is keep that person waiting on hold. If they call to resolve, it's up to the center to allocate the resources to make the call shorter and deal with the customer quickly, before the opportunity is lost.

That's what the hold queue is all about; it's a delicate balancing point between the call center's cost of doing business and the opportunity presented by the customer. Too few resources allocated, and the opportunity will go away. Too many resources and the opportunity may not be worth pursuing. Like everything else about running a call center, how you handle callers on hold depends on the relationship between the urge to reduce costs and the need to drive revenue.

Back to messages on hold. There are two ways to use hold media. One is what I will call the opportunistic method. That involves telling people more about your company and products you offer in short ten second bursts.

There is little else you can call these but commercials. They are most prevalent on queues that are inbound telesales, and they can be said to help the interaction somewhat by providing cross-sell and up-sell information at no cost. You see this a lot less on the service side. Frankly, people do not enjoy that medium when they call to solve a problem.

In fact, it's a bit of a vicious circle. Because when you allocate resources you are almost always going to tolerate a longer hold time on non-revenue-generating service lines than on sales lines. So when you play advertising messages, they cycle through more often during a typical hold session, and the annoyance factor is increased.

You end up diminishing your returns quite quickly. That's why most of the service applications I've seen tend to use hold media somewhat differently.

They accept the premise that you're going to have to wait longer for service than you are for sales. That service from an agent is somewhat more precious a resource than service from the Web, or an IVR system.

So it makes sense to give people as much information during the call that might either shorten, or complete, that call. Before it hits an agent.

In a true technical support center, for example, there would likely be a series of selection criteria applied before the caller ever went into the hold queue. These can be identifiers, telling the routing mechanism who the caller is and whether there should be any special priority assigned. (This is the Platinum card holder example just mentioned.)

It may also include any special information gathered, like an account number or the number of a service contract, that will give the automated router some classification for the kind of problem that might be coming.

The simplest implementation of this kind of thing is "press one for Windows, press two for Mac," but it can be used a lot more artfully to get the call to precisely the right person, and therefore reduce both hold time and the amount of agent transferring that the call would require.

You can also play messages during the queue that alert the caller to the true length of that queue. If you tell a person that there are 20 calls ahead of him, or that the average call takes 4 minutes, or that there's an estimated 15 minute wait before reaching an agent, that's power in the hands of the customer.

That means you won't have to spend 30 seconds mollifying someone who's waited too long before solving his problem. You can also invite the caller to use other options either while waiting, or instead of waiting. Giving out a Web address is a good idea. Offering a faxback system is also good, but less likely to be used during the wait.

And offering to let someone pass the time in an interactive diagnostic voice response system without losing place in the queue can be valuable on your end as well as theirs.

Because if you are CTI-capable, you can pump all the information about what the caller has queried and heard about right to the agent's desktop, saving time. Prevents repetition, and gives the rep more to work with when diagnosing a problem.

In all, how you handle your hold queue will say a lot to your customers about how you feel about them. It presents a lot of opportunities, and can often be a neglected component of call center operations.

■ Messaging Systems

With a message designed for play on hold, you can do more than keep your customers occupied and your company free from fines.

A survey conducted by Nationwide Insurance showed that messages on hold reduced hangups by 50% to 80% and increased the amount of time callers were willing to spend on hold 15% to 35%. The survey also concluded that sales increased 15% to 35% because of information callers heard on hold. Some of the things on-hold messaging can do for your business:

· Increase sales. You can deliver quick bursts of information about who you are and what you do: products, services, special promotions. You can reinforce the messages in your advertising, driving up sales and increasing revenue.

· Cross-sell. Just because someone calls for service doesn't mean they're not open to buying. Cable companies learned this lesson a long time ago, putting messages for premium services and pay-per-view events on the hold queues for their services lines. Another idea is to promote products loyal customers might not have heard about, like extended warranties and service contracts, or limited-time offers.

· Image enhancement. Face it, there's nothing like silence to tell the caller you don't care much for them. Messaging on hold leaves the customer with a more positive impression of your company. And it can make you seem larger and more successful than you really are, an advantage a small startup might need in a competitive industry.

· Answer frequently-asked questions. There are two kinds of questions you can deal with in your messages. The first is the kind that drive your receptionists crazy: your address, business hours, or fax number. The second is to tell the caller what she is going to need to do to complete the call when they reach an agent. "Please have your account number handy, as well as the model number of the product you are having trouble with."

The hardware you need to implement a message-on-hold system is pretty cut and dried — and not terribly expensive. (Over time you'll spend more for programming and production than you will for the playback technology.) Better still, in the last few years it has migrated from analog tape-based systems to digital players and analog/digital hybrids. That has reduced wear, improved sound quality and made systems far easier to update and maintain.

Most phone systems provide an audio input for you to connect a playback system. Many automated attendant and voice mail systems do, too. There are several kinds of on-hold message players. Which you choose will depend on the quality of the message you can tolerate, and the kinds of features you want.

Analog loop-tape players. These are simple cassette players whose tapes run in continuous endless circles. *Don't buy this.* Technology has completely superceded this option. The more you play a tape, the worse the sound. Continuous loop tapes run all the time, by definition. You'll constantly be replacing the tapes.

Auto-download digital announcers are the next step up. These are cassette players coupled to digital playback systems. The units download the program material from standard tape cassettes into the digital players RAM. Because it gets played back digitally, the endless looping is all electronic — no degradation of the analog tape.

One advantage to this type of system is that cassettes are a very convenient way for production houses to distribute programming. These units are often favored as part of a package deal by turnkey producers of on-hold messaging for that reason.

The digital players themselves start with as little as two minutes of playback capacity. Adding memory adds playback time. Features that add value: re-downloading message from cassette after a power failure (eliminating the need for UPS protection), and at the higher end, the ability to check the message integrity in RAM and re-download if it finds errors.

To make programming easy, make sure you get a system that has a PC interface for system configuration and backup.

Remote-programmable digital announcers. This is the new bread and butter of the industry. Like their cousins, these are RAM-based digital players that are equipped with built-in modems. These let you download program content from a remote computer, or from an off-site producer of programming.

If you have a program that requires constant updating, this can automate the process. Most feature battery backup of RAM in case of power failure.

CD-based systems are increasingly popular, especially in larger call centers. These are essentially programmable CD-players. (It is important to remember that nearly all the on-hold technology comes from the audio world, not the telephony or telecom world. And not even from much of the computer world.)

The simplest CD system lets you cue up a selection of audio tracks from a CD. Your on-hold program supplier may distribute your program on a disc, especially if it's custom content. Also, discs are available that contain generic licensed music, free of copyright and license fees.

And the inevitable next step up the ladder: blending the output from the CD player (the music source) with output from a straight digital announcer (custom voiceover content). You can record a message from your handset, phone it in, change it ten times a day if you like. And the music that sits underneath it is of top quality.

Another way to do it is with the features in the switch itself. Most major ACDs now let you play messages to the caller while in the hold queue; in some cases you can literally program in hundreds of messages, from either local or remote content providers.

One of the advantages to this approach is that messaging becomes just one possibility for callers in queue. With the Aspect system, for example, callers can access other parts of the system (perhaps in response to something they hear in a message), and then return to the queue in their exact spot.

One nice feature that's become more common: because it can talk directly to the switch, some announcement systems can tell the caller how much wait time to expect. You can have a set of messages precreated for different wait-times: very soon (seconds), one minute, two minutes, and so on. These can be interspersed with other custom messages to keep that caller from hanging up.

In addition to these options, there are specialty products that are beginning to make their presence felt, especially at the smaller end of the market, in micro-centers and home offices. Typical PCs come equipped with a wide variety of audio/telephony/modem options, that often combine a sound card, voice mail, on-hold messaging and Caller ID detection in one (often cumbersome) piece of software. These are threatening to turn music and messaging on hold into something as common as the home answering machine. (This, thanks in part to the increasingly shrinkwrapped availability of computer telephony technology.)

But that will only accelerate the process already under way of making your choice one of programming, rather than hardware. Choosing a content producer is far harder (and more important) than picking a hardware platform.

Picking the Content
Most of the companies that provide on-hold services (as opposed to the equipment) are experienced audio production houses, many sporting their own studios, voice talent, scriptwriting staff, and message crafting

know-how. Many are from the advertising world, so they know what kind of information and presentation works in holding a customer's attention in the short bursts of an on-hold queue.

Some operate subscription services, letting you change your message at frequent intervals. If you go this route, you'll want one of the message players that lets you swap in a new message quickly, either by replacing the CD or tape, or through an automatic modem download.

In fact, since many production companies act as turnkey providers, you can get the hardware integrated with the type of programming service you need. Service companies often distribute the hardware made by major vendors.

One advantage of using a service is that you can have them do all your recorded systems: voice mail, audiotext, even IVR prompts, all in a consistent voice. Your company will present a consistent face (or at least, voice) to the world.

There are some things you should look for in a service provider:

1. See how long the company has been in business. It's pretty easy to see if someone is established in the industry, or if they're a fly-by-night operation.

2. Call their clients, ask them questions, and most important, listen to their messages. If you call a reference and find yourself listening to a scratchy, repetitive, or worst of all, boring message, you and that production company are not a good match.

3. Try a reasonably priced package of regular changes and updates. Customers call repeatedly over the years. Hearing the same old message can turn them off, especially if you're not mixing in some marketing or sales information.

4. Communicate your promotions and marketing strategies to the on-hold provider. Give them something creative to work with, and something to earn their pay.

Some companies provide an 800 number to call for a sample message on hold, others provide demo tapes. Either one will be valuable in examining what the company has to offer.

The more "creative" you want the company to be, the more important listening to a program similar to the one you have in mind is. Have your heart set on a humorous program? Listen first. One persons humorous is another persons corny.

(Humor is not the universal language, this may be why so few companies use it in their messages on hold. If you are interested in a humorous message though, many providers will be eager to do your bidding.)

Crafting an Effective Message

Here are some tips for what to put on your on-hold messaging system — pointers that should keep people holding longer.

1. Don't keep telling customers to continue to hold. Customers find this irritating. Use the time efficiently to promote.

2. Give callers an option to exit the message. The customer needs an out. They need an option to leave a voice mail message or reach the reception area, especially in the call center environment where they may want to place an order or voice a complaint. Callers don't like being forced to listen to a message and if they have to, there will be more abandons.

3. Steer clear of chest beating. Keep the message short, low-key and to the point. The message should be informational rather than advertisy. The customer wants to be taken care of, not hear bragging about the company. It should be viewed as an opportunity to tell about products and services, offer an toll-free number for future calls or a fax number.

4. Use MOH as an opportunity to tell customers about other products. Product A might sell by the thousands, but customers may not know about B and C products. Using the phone to tell them is a critical link. There's already a relationship. It's been proven that people will do repeat business with a company they're comfortable with.

5. Incorporate thematic elements into an on hold message. Tie in mottos or slogans used in brochures or even yellow page ads as part of a complete marketing mix.

Give your audio program suppliers an activity schedule of what you're planning to do over the year. That way a tape could be in and running for whatever you're marketing that month.

6. Watch for excessive repetitiveness. The first and second time a caller hears the same message is okay, but once they hear the same message on a third call (or the third time within a call) you're in the danger zone.

There's no excuse for not using some form of on-hold messaging. Its a cheap, easy way to keep people from hanging up. You can have someone else do nearly all the work, while you reap the benefits. How often does that happen in life?

QUICK TIPS

1. Writing your own scripts? Some pointers:
   - Keep individual topic items short.
   - Thank the caller often.
   - Use phonetic spellings for unfamiliar words.
   - Don't make any one topic more than four lines long.
   - Don't use "shop talk."
2. Don't succumb to the temptation to make your message a no-holds-barred extravaganza, catering to every caller and using every auditory bell and whistle. Messages should be simple, to the point and NOT ANNOYING. That's what will keep callers on the line.
3. You might be thinking: Why should I invest in programming or production? Can't I just play the radio on hold to keep people amused until an agent is available? The answer is: NO. Playing the radio over the phone amounts to a rebroadcasting of copyrighted material. You can buy a license in advance (for a sum that depends on the number of phone lines you have). If you don't do it in advance, you can end up owing ASCAP and BMI much more, a fee of from hundreds to thousands of dollars per song. And, if your competitors find out you are playing radio on hold, where do you think they'll place their ads? Finally, you'll lose all the benefits of having someone call — you're not selling to them, or giving them some value-added information that's going to gain (or keep) a customer. And you're probably paying for the call, too. Think about it.

Part Six

Management & Operations

Call centers operate under two
contradictory mandates: they are
charged with improving service
(or sales, or revenue), and at
the same time, they are under
pressure to reduce (or at
least cap) the costs.

Workforce Management Software

Estimates vary, but all reports I've seen show a steady growth in the call center industry that's been going on for the better part of ten years now. As the industry grows, the skills involved in managing centers have become their own specialty; being a call center manager is now something portable, that a person can take with them from company to company, even as that person changes jobs across industries.

At the same time, the number of people employed in call centers has reached staggering heights. Some estimates (rumored to come out the US Labor Department) say that perhaps two to three million people in the US work in call centers.

By any measure, it's an industry that's saturated, and that has a historically tough time attracting a steady supply of acceptable workers.

However. The flip side of that very good news includes some down side. Labor is very expensive, and in the US at least, very hard to keep and maintain. A benchmarking study done by the Purdue University Center for Customer-Driven Quality found that turnover is a real problem industry-wide (not just in your center). In fact, inbound centers have an average annual turnover of 26% for full-time reps, and 33% for part-timers. Nearly half of centers said that part-timers handle 5% or less of their total calls.

The study broke apart the figures for full time and part time reps, and had very few surprises but lots of hard data on the subject — highly recommended for info on lots of call center metrics.

Of course, high levels of turnover lead to higher training costs, which in turn put pressure on the centers overall productivity.

In some sense, call centers are operating under two contradictory mandates: they are charged with improving service (or sales, or revenue, or whatever desirable metric), and at the same time, they are under pressure to reduce (or at least cap) the costs. If done right, this dynamic can lead a center and its company to be more efficient and more productive on all kinds of levels. But the fact is that the cost of hiring & training reps needs to be brought under control if a call center is going to actually be the engine of company productivity and regional job creation that people expect it to be.

Where am I going with this? To the amazing fact that workforce management software is found in only about 10% of call centers.

Workforce management is the art and science of having the right number of agents at the right times, in their seats, to answer an accurately forecasted volume of incoming calls at the service level you desire. Naturally there is a whole class of software that accomplishes this task, much of it quite good.

In fact, the whole notion of being able to accurately predict call volume and then staff accordingly is so tantalizing that it almost makes you wonder what the problem is? Why are so few centers using it?

Let's caveat that first: it's likely that the 10% that are using it are among the most advanced, highest volume, highest technology, and productive of centers.

But it really does beg the question, if it works for them, why are so few of the rest of centers actively using this technology?

The answers are a little murky, and not always verifiable by cold hard fact and market research. Here are some of the thoughts:

· WFM is thought of as a very high cost endeavor. And it's largely true. Systems that predict call volume and match staff schedules to that volume historically cost a lot of money. It could take anywhere from $50,000 to $100,000 and up for a really advanced system, especially one that spans multiple sites.

· They're also thought of as high-maintenance babies. There's a perception out there in the world that in order to run a fully configured workforce management system, a call center is going to have to have someone on hand, making schedules, feeding data in, going over the data that comes out, literally babysitting the system at the cost of 3-4 agents, maybe more.

This perception is sometimes reality. The more complex the system, the more training it's going to take to run it, especially when you are scheduling and predicting across multiple sites.

· There are "cultural" barriers to a greater market penetration. By cultural, I mean that call centers have traditionally put more emphasis on managing the call, and its flow through the system, than on managing the workforce. This shows up in a historically high turnover problem (there's that again), and something of a disdain for the workforce that comes up every time labor and management talk about things like monitoring.

So there are many call center managers who are simply more comfortable putting schedules together like the pieces of a mental puzzle, or who can't see the justification in an expensive software package. These are valid points of view, until someone comes along and demonstrates the return on investment workforce software can deliver, which brings me to the next point:

· Lack of vision on the part of the vendors. The companies that make this software have not done a terrific job articulating to the other 90% of the market their vision of the benefits. Sometimes that comes from complacency, as when a market leader doesn't see the need to bang the drum for a technology. Other times it's from basic incompetence, as in the case of one particular mismanaged vendor that bought up a WFM company and then didn't know how to either improve the product or sell it to customers, and in the process wrecked it.

· Finally, there is a disjunction between the seeming simplicity of creating a schedule, and the real complexity it takes to make it the best of all possible schedules.

The typical call center manager has a range of options for creating a schedule. He or she can cobble it together from back of the envelope calculations. Or create formulas in a simple spreadsheet. Or buy one of the $500 Erlang calculators that let you input your center's variables. And so on, all the way up to the 5- or 6-figure full-fledged programs. But to do something really productive really does require some powerful software, and that's going to cost money.

▣ The Benefits

The main benefit, of course, is more efficient scheduling. This should go without saying, but I have to say it. More efficient scheduling *saves you money*.

You can be prepared for sudden changes in call volume, those annoying peaks and valleys that come along seemingly without warning. Here's the take-away: workforce management gives you warning. This software is so good, it can see the patterns in your call histories and discern the peaks and valleys that even the experienced call center manager won't be able to see coming.

Holiday scheduling is a good example. Holidays are when the two elements that most directly affect the call center go into a sudden divergence of interests. Calls surge up in crazy ways. But they are predictably crazy, if you know the patterns that drive them. People, on the other hand, tend to want all sorts of counterproductive things, like days off, flexible schedules, vacations, time with families. Kids are home from school, and so on.

Workforce management software allows you to project the call load for a given day from the historical data. You'll know at any given moment how many calls are going to come in, and you can match that most effectively to those human resources available, even in crazy times. You can act quickly to handle any divergence between people and calls, either days ahead of time, or within a shift.

This is just a small example of the efficiency and optimization it brings. Everyone will have to do their own ROI calculations, of course, based on internal factors, but the broad outlines are almost always going to show an improvement in service level and call handling capabilities. When exploded to take into account the harder variables that are affected by worker morale, like turnover, training costs, it will look even better.

And you can also create custom schedules for people with special needs, like students, older people, parents, and people with medical problems.

There are other good things about workforce management software:

Threshold alerts. Supervisors have instant information about intrashift variations that could cascade through the day and cause problems later. They can then adjust schedules on the fly.

Performance evaluation reporting. Take all the data coming from the ACD and make it meaningful. Workforce management is not the only place to collect this data, but it can be one of the most meaningful. Having a coordinated real time and historical (short-term) view of activity is better than spreading that information, and its analysis, out among different software tools. This creates islands of information that are harder to put back together later.

Learn why you're not meeting service levels. Answer baffling questions like: was an entire group's abandon rate higher because someone took lunch a little too early, or because you had too few people on hand for an expected spike? Or, did you have too many people, and have a perfect service level at an unacceptably high cost? And, what would happen if you added ten people to that shift? Don't pay overtime for people you don't need. And have the justification calculations available when you really need to make the case for more people.

Coordinate between sites. Pulling agents from multiple sites together as one virtual center gives you all the workforce efficiencies you'd get within a single center, with greater economies of scale.

Put power and information into the agents' hands. Schedules can be worked out in ways that let the agents understand the why's and how's of your decision making. When there's hard prediction about call volume, and an automated scheduler that optimizes everyone's break and time-off preferences fairly, there are fewer reasons for griping and complaint.

Simulate conditions and changes. Workforce management, when combined with simulation software, takes existing (or historical) conditions and lets you tweak the parameters to see the what-would-happen-if? scenario. You can see the effects of adding people, subtracting them, changing group dynamics, even of adding different technologies (like an IVR system) to the front-end.

QUICK TIPS

1. Simulate conditions: Do you ever wonder what would happen if you let two agents go on vacation at once? Or if you added a part-time agent for a few hours on Mondays? Or what would happen if call volume increased? Using software to create these what-if scenarios lets you know how high abandons will shoot up and how long callers will be likely to wait in queue. Find out the effects before you make the changes.

2. Some centers have supervisors and managers walk around and tell agents the general status of queues and wait times. When things get bad they run around and yell at agents. Using readerboards to pass supervisor statistics on to agents increases productivity. It gives agents a chance to monitor themselves and decide when it may be okay to log out for breaks and lunch. And it gives supervisors freedom to move about the center, giving individualized assistance.

3. Make the most of your reports. Try segmenting agents into work groups based on similar salary levels and other attributes so you can compare how each agent is performing relative to others in the work group.

4. A reader recently asked me whether I could think of any practical measures for performance and optimization beyond the usual "service level" and "calls per head." In general, call centers should tackle optimization and metric questions based on a cool assessment of how the center relates to the rest of the company, and what the company expects from that center (vis a vis competitive pressure from the rest of the industry). That is, airline call centers measure different things than catalog order-takers, and that's fine.

   You might want to think in terms of results that impact on what your call center is actually trying to accomplish, and how that impacts revenue: call duration, for example, can impact both costs (telecom transmission charges) and customer satisfaction (if you're using the time to sell the caller some new offering).

   You can measure center performance as a whole by using a workforce management system and keeping track of schedule adherence; the closer you are to the predicted schedule, the more optimally you've staffed. It helps you keep costs from ballooning out of control unexpectedly.

   And you can measure the performance of individual agents and groups by tying their metrics to actual customer information. (This requires some CTI and/or back-end integration with customer data.) What I mean is, you might be able to generate a revenue figure for each group or rep that weighs their call length or number of calls taken by how valuable those calls are. An agent that handles fewer calls, but calls from premium customers with a high lifetime value to the company, is probably more effective than an agent that handles more calls in shorter time, but where those calls are from low-impact customers, or not customers at all. This will depend on internal factors within your own company and your industry.

■ Competitive Advantages

So what does this do for you?

First of all, it's better to *Know* than *Guess*. To have a handle on costs, calls and variables is to have the power to react to changing circumstances.

Second, it enables you to have a better managed, more informed, happier workforce. This makes you more attractive as an employer, and makes it easier to attract reps in regions where workers are hard to get, and hard to keep.

Third, penetration of this software tool is so low, that the simple act of installing workforce management software automatically confers a technology advantage on your over your competition. You force them to do the same.

I think it's fair to say that it's one of the least used, and most useful, technologies available to the modern call center.

It may not be as sexy as something like Web/call center combos, or voice over IP agents, but it works, works well, works as advertised (important when so many call center technologies now are so much vapor), and it can save a lot of money and heartache.

Monitoring Systems

You wouldn't train an agent without listening to his or her phone technique, would you? Monitoring is a critical part of the process of teaching a new rep how to deal with customers, how to handle difficult situations, even simply how to follow a script and read a screen full of complex information.

Feedback is important. Without it, reps don't improve. Even reps that have been on the phones for some time need constant skills assessment and further training. That's just common sense.

In most centers, you come at quality from two directions:

· making sure you have the best agents possible, operating at the highest level they can personally achieve; and

· enforcing a consistent standard of quality for customer contacts, from the customer's point of view.

Monitoring agents is still the best way to ensure that you achieve quality from both standpoints. If handled with sensitivity, monitoring can be a benefit to agents because it helps them define and reach career goals, assess strengths and weaknesses, and mark their progress according to realistic standards.

Opponents of monitoring cite its possible negative effects, like stress and reduction of employee privacy. But in products that monitor agents, the emphasis is on how monitoring can help both agent and manager work together to improve performance.

Pros and Cons

It's possible that you have a monitoring system built right into the switch. Some manufacturers, like Teknekron build in sophisticated software for combined monitoring/quality assurance programs.

E-Talk's *AutoQuality* and *P&Q Review*, for example, are tools for collecting data about agent performance and assessing that data over the short or long term. *AutoQuality* automates the scheduling of agent monitoring for later review. Managers don't need to be present to monitor, or to set up tapes.

Many headset vendors have training-based models available that add a second jack on the amplifier, to accommodate a no-mic headset that, presumably, the trainer could wear when sitting next to the trainee. Or, as a really low-budget monitoring system, you could plug a tape recorder into the jack.

The most obvious benefit of monitoring is that, if done right, it creates an objective standard of behavior that can be measured, and if found good, repeated. It helps ensure that you deliver not only good service, but also consistent service. Consistent from each agent, and consistent across agents.

From the agent's point of view, it creates a way to measure performance that can be spelled out in advance and critiqued intelligently. Results can be quantified and reps can see improvement over time. Plus, it allows management to benchmark standards and ensure that all reps are treated fairly, by the same standards.

Tools to Use

So what's out there? This is a surprisingly robust area, with technology advancing rather more rapidly than you might think. For one thing, the growth of the Internet is making it easier to store and retrieve information across networks, even if that information is an audio recording of a call.

Training is one of those things that everyone knows has to be done, and most make an effort to do it. But in call centers it's hard to pull people off the phones and put them in classrooms and then later, to measure whether the training is having any effect at all. Training needs to be targeted to the needs of the individual, and it needs to be done where and when the agent has time.

One way to enhance the skills of agents is to turn to the information gathered through monitoring tools to tell you who needs help, and of what kind. Monitoring has evolved beyond simple recording of voice and even data. It's becoming a vehicle for sophisticated analysis of needs — both individual, and in the call center as a whole.

Witness Systems has something that delivers ongoing training that's tailored directly to the needs of specific agents, and to their desktops. *EQuality Now* integrates with Witness's main *eQuality* system, which performs the recording and archiving of customer interaction information. This new tool is an extension that takes the information about who needs training, and prioritizes bite-sized modules of training ("chunks") that are accessible through browser-based desktops.

In essence, the system analyzes performance to determine what skills a particular agent needs to work on to improve overall performance. The center has to create modules of training, but once created, they can be delivered to any rep at any time, and it's always a consistent bundle of information. Results can be measured, analyzed across different reps, and across the performance of a single rep over time.

One of the things I like about this is that it doesn't discriminate in favor of telephony interactions (aka "voice calls"). It starts from the premise that if you have a pool of reps whose skills are based on answering calls, and you ask them to start handling emails and Web-chats, you're going to have some performance and training issues to deal with. Your chunked training modules can be geared to improving any set of skills — but the system is built to handle the monitoring and analysis of any of those alternate interaction channels.

In a world where turnover is extraordinary and the labor market is white hot, being able to deliver appropriate training will certainly help keep reps, and prepare them for the demands of the multichannel call center.

This kind of tool ensures that training isn't going to be an afterthought — *eQuality Now* is part scheduler (so you don't have to worry that your call center will be empty with everyone in the conference room when the phones start ringing) and part delivery system. And moving up the chain, it connects to higher level CRM tools that can give you a broader picture of how well the training (and by extension, the center) is meeting corporate goals.

Security verification and "quality assurance" are not enough of a justification for a lot of call center buyers to put in a system that employees find intrusive and that requires a lot of connectivity and tech to make work well.

Where the recording companies are taking us, though, is past monitoring. They are making the very logical argument that when you look beyond the simple recording technologies that capture calls, you find that they are able to collect an enormous amount of data that's ripe for analysis. And when you analyze it, you can do some very useful things with it: pinpoint who needs training, and get it to them, fast.

The idea is to change the way we think about monitoring. Instead of having reps go through the day thinking that every call has be perfect because it might be the ONE call that the supervisor is listening to, the rep should be almost oblivious to the recording. The information collected is, now, more important in aggregate, and when analyzed by machine, can show patterns that a listening supervisor might now detect. Then you can direct the person to the kinds of training that make the most sense. In a hurry.

The kinds of data that are now routinely captured along with the audio portion — things like the agent's screen activity, or the Web page that the caller was looking at when he completed the transaction — are combined to bring a new kind of detail to the verification and quality monitoring process.

On the other hand, sometimes monitoring tools can be less than helpful. As if call centers don't have enough trouble with turnover, an Israeli company called Trustech has come out with a product called *i-Check*, a voice analyzer that dynamically analyzes the speech flow of either the agent or the customer, during a call.

That tells supervisors exactly how the agents are "feeling" during the call. They represent an agent's "feelings" by reporting on stress levels and other psychological indices. The theory being that this could then be used to enhance the management of customer relations within the call center. They apparently perceive that this could be used in conjunction with a monitoring application that stores calls, retrieves them on demand and runs them through the analyzer. It includes a suite of tools that includes both real-time and off-line stress diagnoses.

But there's something really creepy about a product that detects stress in an agent's voice and alerts supervisors. The CEO of Trustech said in a statement that this product "greatly enhances the agent's performance." I think that's hooey.

You think you have problems with agents and monitoring? Try telling people that their performance is going to be evaluated based on their emotional levels and their stress. Is there any agent who doesn't experience stress? There are so many better ways to measure performance and reduce tension in the call center workplace. Let's stipulate that reps get stressed. Many of them even hate their jobs. But the quickest way to turn the call center into a high tech sweatshop is to give them a numerical score based on how stressed out they are.

What To Do

So in thinking about monitoring, and how it can be used well (and badly), I've come to some perhaps obvious conclusions.

Recording technology, which has been around since dinosaurs walked the earth, has been improved twice in the last decade. First, when digital recording replaced analog. (This makes it easier to store, and then later retrieve, specific calls.)

Second, when CTI links made it possible to turn that digital recording into a piece of data, and marry it to other information about the transaction: screen scrape.

While those things fascinate, the debate about monitoring and its effectiveness goes on. Because sometimes it's easy to forget that monitoring is about judging people and their performance. And recording technology by itself is not a sufficient tool for making informed judgements.

One tool I like is Witness Systems' *Performance Analyzer*, software that works in tandem with core recording systems to measure agent performance against specific goals. *Performance Analyzer* helps solve the problem of accessing disparate information throughout the enterprise by serving as a central repository for information from many sources, such as workforce management, human resources, predictive dialers and ACDs.

Combining, assessing and exploring information from multiple sources is critical to today's evolving contact center because no one source has sufficient information to provide a complete performance picture.

Previous tools have concentrated on analyzing just the data that comes out of the recording system; this, more broad-based tool lets you combine streams, let's you look at performance trends from a variety of perspectives.

And unlike earlier tools, this lets you evaluate performance of agents or groups. Or you can scale that analysis up the chain to look at an entire center, or group of centers. Add information from accounts receivable, order entry, and the like; then you have a picture that tells you more than just: "Did this agent follow the script when we randomly recorded him?" It can tell you how much money that agent generates, and whether a particular campaign is in trouble.

Simple monitoring, though useful, needs to be mitigated by data, by thorough, rigorous — and fair — data analysis tools like this one. The successful monitoring plan is one that can gather all the dangling threads of data that pass through and around a call center and turn it into a rich tapestry of useful information.

Often, when people speak of finding patterns within call center-related data, it's based on someone outside the call center sifting through the customer data records, usually from a marketing point of view.

But what if you could take all that customer data and marry it with two other important streams — the performance data from a monitoring system, and the call detail information from the ACD? Then you'd have a really useful pile of data, if only you could make sense of it. That's where the data mining tools come in.

Especially since most of the available monitoring tools now include some capability for "screen scrape," or gathering the data that passes through the agent desktop application during the call.

Let's stipulate that data mining is a good idea. Reasonable people can argue over the merits. What I want to get to is the potentially awesome mind-shift that that puts into the center, which is this: for it to work well, you literally have to record every call.

In practice, that's not so different from common random monitoring; most people will set up their system to record a certain percentage or number of calls at random intervals, and of those recorded calls will use some or all for evaluation purposes. There are, of course, some applications that already require total recording for auditing purposes, but those are pretty specialized. Indeed, in the typical call center, the notion of total monitoring is something of a new idea.

Storage is nearly limitless and inexpensive now, so that's not an issue. Screen scrape poses a different problem — the limiting factor to total recording is not storage space, but LAN bandwidth. Pushing screen scrape beyond random causes degradation of the responsiveness of anything else travelling over that LAN, including the agent desktop apps.

So how will people react to the idea of total monitoring? Is there going to be a discernable agent response to something that's more pervasive and less clearly based on quality assurance and performance evaluation factors?

Monitoring is a lot less controversial than it used to be. I think that's because of the increased use of "fairness" tools — automated systems that are used to ensure random recording and calibrated scoring of agent performance. Those tools take a lot of the sting out of having a supervisor listening in over your shoulder. We also have progressed quite far from the days when monitoring was literally an "over your shoulder" experience, with the supervisor jacking a headset right into the agent console and listening during the interaction. If anything could ensure a jittery and atypical interaction, it was that.

You can make a pretty fair case that total monitoring is just as fair as random recording. You can't argue that you're being singled out if it's happening everywhere all at once.

It still needs to be explained to agents, however, that total recording doesn't mean total evaluation of every call, that there are still random factors involved in the selection.

■ Benefits & Risks

With that, it seems a good time to reiterate the pros and cons of monitoring in the first place. The benefits are these:

1. You create standards of performance, and a way to assess whether those standards are being met. Not only does this give you better measurements of your workforce, but it tells you if your standards are absurdly high and therefore unattainable.

2. It allows you to correlate training regimens with those people who actu-

ally need it. It also helps you build teams by discerning patterns of complementary skills among reps of different abilities.

3. It creates an audit trail for the customer record, allowing you to see exactly how an interaction was handled and find patterns in how customer interactions go off the rails.

4. Automated, random monitoring ensures that no agent gets an unfair burden of monitoring, and it ensures that promotions and incentives are doled out according to impartial measurements.

5. Capturing the screen of the agent as well as recording the audio part of a call shows you more about what that agent knows and how he or she follows procedures.

6. And as noted, technology now coming available lets you search larger pools of data (coordinated among audio recordings, screen scrapes, and added ACD data about call durations and transfer pathways) for patterns that show you more about your center as a whole, not just an individual agent's performance.

On the con side, remember this:

1. People do not like monitoring. No matter how you bring it across to the rep, know that monitoring can be seen as intrusive and unfair. All the productivity and audit trail benefits in the world won't make someone like being observed and recorded.

2. It's possible to abuse it by not calibrating the responses and scoring among supervisors.

3. It's not going to be worth much to record calls automatically and then not use one of the many good quality assurance software tools for consistent scoring.

It's encouraging to see all these data pathways coordinated in a way that gives equal weight to the performance of the agent during the actual call in assessing customer satisfaction. People talk about CRM — but that category pays scant attention to the reams of data contributed by even traditional random monitoring.

Making Call Center Careers Meaningful

It's no secret that call center reps are hard to find, and keep, in this white hot labor market. What's not so widely known is that the competition is fierce and keen for call center management professionals, as well. Though there is no data right now on this, it's my belief that this scarcity extends downward into the realm of middle managers.

As far as reps go, it's not hard at all to see why they are so hard to hire and keep. Frankly, in most cases the job leads nowhere. It's notoriously low paying, and has little chance (in the US) of becoming more than a way for entry-level or transient workers to pick up basic technological and customer contact skills.

There is always discussion in this industry of turnover, its causes and effects. The way call centers traditionally deal with agents is to think of them as aggregates, as figures plugged into equations of profit and productivity, and always measured against that most dreaded and misused yardstick: the call.

The fact that the industry simultaneously needs more managers, and more staff, are of course related. There are two broad things that the contact center business needs to do to have a better future. But first, a tiny bit of history.

The farther back into the past you go (up to about a quarter century or so) the more the call center looks less like a single industry with a coherent structure, and more like a collection of isolated and varied ways of performing certain customer-related tasks for companies in well-defined vertical industry sectors. A call center for a bank, for example, would be run by someone who had a great deal of experience with banking tools. This person would consider himself (in those days it was usually a "he") a bank professional. If he went looking for another job, it probably wouldn't be managing a center for a cataloguer or a health care firm. It would be for another bank. And the job would have less to do with "running the center" than with managing the technology, especially the data.

In fact, knowledge of either the company's proprietary IT structures and data models, or the ability to oversee raw telecom, were the two criteria for success as a manager in most corporate call centers of that day.

Over time, the many isolated islands of call center activity began to recognize that there were commonalities across industries. Now, it's taken as an article of faith that a catalog retailer is not much different from a bank or a software company, in so far as the call center will look pretty much the same in form and function. This, above all, was the revolution that took place in the 1990s: the recognition that there was such a thing as a call center industry, and that it had less to do with "telemarketing," which nobody liked, and everything to do with "customer contact," which excited everybody.

Once a call center professional class was recognized, ideas and managers could flow freely across sector boundaries, and the result is the huge business we have today.

But the nagging question still remains: what makes a manager? How do you determine criteria for success, so that you could write it into a job description and hire the best person?

That leads me to the first thing that needs to happen to ensure continued success: we need to admit, once and for all, that knowledge of technology and its structures are an imperfect and outdated selection criteria for a call center manager. Admit it to yourselves now: it's all about people skills.

Being able to corral enough people into seats not just today, but to have the vision of how to ensure a steady base of MOTIVATED workers for the entire five to ten year life cycle of a call center; to be able to anticipate the way changing technologies will affect the mix of ways customers will be dealt with, and ensure that the call center remains the nexus for contact, and remains of continuous value to the organization. The old way of hiring a manager was to make sure the person had experience, and knew his or her way around the switch and the CTI software. That ensures that a center will remain a static place where "triage" mentality lasts until a center is rendered irrelevant by a shifting customer base and different organizational needs.

What a manager needs to do is provide a company with a base of tools and people for delivering any and all information about customers that may be required. And to anticipate the kinds of information that may be needed.

Too many call centers are looking in the wrong places for the wrong people.

Those readers with a keen sense of what's coming are probably thinking that I'm going to recommend that we start looking at the agent pool as a place from which to recruit managerial talent. Far from it. But the problems are related. And I think that as long as we look at the agent question from the wrong direction, we're not likely to have much luck changing minds about management either.

Here's the second way we have to change things. Agents are currently motivated by a combination of factors and methods, but in the main those things come down to simple productivity. Answer more calls, answer them quicker, process more transactions, bring in more revenue, and in some cases, get better scores in quality assurance systems. It's mostly about doing more, or doing it cheaper, than it is about raising the value of the company in the eyes of the customer. For all the talk about customer relationship management, precious little of that mantra trickles down to the agent's day-to-day interaction with people. Putting a new icon on the agent desktop doesn't make a relationship.

Let's be cognizant of the fact that call center reps are often accorded just slightly more status than the people who pack boxes in the warehouse. Yes, it's a transient and often unreliable population. It's often under-educated. It needs constant training in skills for dealing with machines and with people. But few call centers see it as an incubator of ideas, and that's sad.

Fewer still see the call enter as their eyes and ears to the wider world, and that's not so much sad as just pathetic. The call center business is full of jargony notions, like this one: the call center is "your front door," that it's a "single point of contact," and so on *ad nauseum*. In fact, the call center is a place where disposal people are used as cannon fodder to absorb the first round of customer difficulties. This is a waste of resources, and of course it's a wasted opportunity to do more business.

What you must do is see the agent as the chance to gather intelligence on customers, and even on competitors. If you want to keep agents, you must pay them more. You must tell them that yes, this is the grittiest job the company has to offer — the job where they are going to learn the truth about what customers think of the company. There is, as we all know, no harsher truth. If you want to keep agents, you have to tell them that what they learn has a point: to benefit the company. And that what they learn and contribute will pay off in a better company, and ultimately in a better job. They have to be encouraged to look beyond the call center for opportunities. And they have to be encouraged to look upward for chances to grow. Not just to "senior agent," or "supervisor," but to something real and long term.

Remember the apocryphal tale of the hard-as-nails CEO who worked his way up from the mail room and succeeded because he knew more about the company than anyone else? Someday there's going to be a CEO who worked herself up from the call center, and succeeded for the same reason.

■ Turnover Affects Corporate Performance

Want concrete evidence that the way you manage your front line call center staff has ramifications far beyond the call center? A recent study showed that employee turnover replacement costs have reduced earnings and stock prices by an average of 38% in four high turnover industries, including call center services. That's right: stock prices and earnings.

Research by an operating unit of Nextera Enterprises, a management consulting firm, shows that turnover rates ranged from 31% annually in call centers to 123% in the fast food industry. The research analysts estimate employee turnover is costing companies in these industries more than $75 billion to just replace the more than 6.5 million employees who leave companies every year. Average turnover across all US industries has climbed to nearly 15%.

"Employee turnover is draining profitability from companies in many industries," said Jude Rich, chairman of Sibson & Company, the Nextera subsidiary whose talent management practice conducted the study. "By reducing turnover, the opportunity to improve a company's stock price can be substantial, but many companies have not declared an all-out war against turnover. I believe that there are several reasons for this: many managers do not know how much turnover really costs; others have not figured out the root causes, so they do not know what actions to take; while others mistakenly believe turnover is inevitable in their industry."

Direct employee replacement costs may be just the tip of the iceberg. "Employee turnover has a significant effect on companies' top lines by inhibiting their ability to keep current customers, acquire new ones, increase productivity and quality, and pursue growth opportunities," said Seymour Burchman, a principal of Sibson.

"At a company that specializes in services within the home," Burchman said, "we found that customer cancellation rates were three times higher for new sales technicians than for experienced ones, principally because customers are more comfortable when they see familiar faces providing services at their homes. Today's tight labor markets have exacerbated the problem, forcing many companies to operate under capacity — managers are having difficulty finding people to replace the workers who leave, much less hiring people to pursue growth initiatives. An employee retained, is a new hire avoided."

The study focused on "front-line employee" turnover (those employees who have direct impact on a company's customers) because "front-line employees can have a significant impact on revenue generation, productivity, and growth potential," said Burchman.

Additionally, Burchman pointed out that the impacts of turnover are not limited to just companies in these high turnover industries. Companies with high turnover rates in other industries are often unknowingly experiencing material reductions in their earnings and stock price because of turnover. "For companies experiencing high turnover, there is a strong potential to increase their stock price by reducing attrition," said Burchman. "This is true in companies in many industries. At a telemarketing service call center, we found that if management halved the turnover, which management agreed was attainable, the company's market value would be increased by one-third."

Knowing what actions to take begins with targeting employee segments with high costs of turnover. At a technology company, 5% of the employees made up 42% of the costs. Once high-cost employee segments are identified, it is important to determine the root causes of turnover for a particular employee population.

Many companies are under-investing in turnover reduction because they don't have the information they need to calculate the ROI. One analyst notes that managers must then implement creative solutions in areas that matter to employees. For example, at one call center, a company put trailers on a college campus to attract students who did not have cars or could not afford the commuting times.

■ Steps to Reduce Turnover

Sibson has found there are four key steps to reduce turnover:

· Make a quantitative, financially driven, business case for change by identifying the total costs of employee turnover, and the potential savings if turnover is reduced.

· Develop a robust qualitative and quantitative fact base, to identify the root causes for turnover of specific employee segments.

· Come up with tailored, creative solutions aimed at eliminating the root causes, rather than the flavor-of-the-month solution, or what another company tried. For example, throwing more cash or benefits at turnover caused by poor advancement opportunities or unacceptable job content will not work.

· Identify the necessary investment to increase employee retention and the expected return on that investment.

Chapter Thirty-One

Surefire Ways to Motivate Your Reps

Getting, and keeping, call center reps is one of the most difficult things to do well consistently. With high turnover almost structurally embedded in the nature of the modern call center, one of the smartest things you can do is figure out exactly what things will motivate your reps, keeping them on the job longer, and keeping them more productive.

Working in a call center is a difficult job. Although working conditions are generally good, reps are often unappreciated, underpaid and not encouraged to aspire to move up the company ladder in any meaningful way.

Many people see the reps as a weak link in the customer/company interaction chain. However, it is important to remember that these are people you have trusted with your most valuable corporate assets: your customers. If you don't have faith in them, or at least do something to show them that you have faith in them, they are going to leave. And you'll be right back where you started, recruiting, rehiring, retraining and extremely frustrated.

Here are some suggestions for creating better working relationships between management and call center reps, as well as between and among teams of reps. Some of these suggestions have to do with certain technological enhancements that can improve productivity and meet the daily life of a rep better. But, as you'll see, some of them are rooted in plain old commonsense.

1. Pay more. Do I have to say this? Yes. It's not unusual for call center reps to make less money than secretaries and other administrative support staff. Call center reps shouldn't be thought of as the support staff. They should be considered professionals. What they do requires training, skill, and a certain amount of grace.

   Now, readers from outside the U.S. must be smiling, because they know that raising U.S.-based wages is going to make people consider outsourcing some or all of their call center operations to emerging call center markets. This is no doubt true. One of the key ways in which those emerging markets compete for business is with the lure of low-cost labor. That does not mean that U.S.-based call centers can be stingy. Not unless they want to continue having difficulty hiring new workers, and dealing with a 35% turnover rate.

2. Extend benefits to part-time reps. Many call centers treat part-timers as less valuable than full-time reps. In many ways this does not make sense. Part-timers are the glue that allows a call center to handle peaks and valleys in volume at odd hours and during critical seasons. Along the same lines as paying fair wages goes extending things like flextime, health insurance, and vacations to part-timers.

3. Allow shift trading between reps. This is one of those small, easily accomplished managerial decisions that goes a long way toward making people happy. It means that someone can come to work without worrying that is sick child or final exam is going to mean joblessness or hardship.

4. Use advanced workforce management software to more easily accommodate the changing schedules of students, single parents, and other "transient" but necessary workers. You'll be able to afford better workers if you are using software that more efficiently schedules reps to work at optimum times. If you have the right number of agents in seats at all times, you'll be able to afford accommodation without busting the budget. Plus, agents are less likely to gripe if they know that special requests are being handled fairly.

5. Create an "off-site" agent program, enabling reps to work from home using their personal computer equipment. So many systems these days enable remote communication to a center through a browser-based PC. With a minimum investment in technology, you are able to take advantage of agents working at home, from small satellite centers, really from anywhere. Allowing reps who excel and who are responsible to have this perk can motivate others to perform well.

6. Create an advancement system, whereby reps know that if they succeed, and add value to the company, they will be rewarded with more responsibility and better pay. You might pay them, for example, to help in training others. You might reward them with specialty training for other areas of the company, or even with trips to industry trade shows and seminars.

But most of all there must be some provision made for people to advance in the company. Reps can advance to become supervisors, perhaps to become managers themselves in time. But the progression must be obvious to the reps.

7. Provide for ongoing, constant refresher training for all reps. This applies even more strongly to the reps that have been there for a while. (Ongoing training may be the best way to show reps that you care whether they stay or go.)

8. Bring experienced reps into the training process so that newer reps can benefit from their expertise. I don't just mean how well they do on the phone; it includes what they know about your company, how it works, how to get things done, cut through red tape and generally apply more human intelligence to the customer call.

9. Use an automated — not entirely manual — monitoring system. Although this sounds counterintuitive, in that the one thing agents tend to resent the most is the intrusion of monitoring, if you make clear from the start that the monitoring is being done in a completely impartial and nonarbitrary manner, with an eye toward quality assurance rather than performance evaluation, it is far more palatable.

10.Consider bringing in spouses. There are few people who know your call center better than the spouses of the people who work there. With a little training those family members can become an inexpensive reserve for those times when you need a few extra bodies in the seats. Real employees will see this as a sign of trust and an opportunity to bring a few extra dollars into the household.

11.Avoid mixing traditional telephony interactions (i.e., "calls") with other kinds of interactions that require a completely different mix of skills. Instead, consider handling those other interaction channels with small dedicated groups of specially skilled reps. Not only does this better match the skills of the rep to the kind of work being done, but it gives reps another point of view from which to examine their work. For example, a rep whose performance is poor, or whose skills are poorly matched to handling customer interactions by telephone might find a fuller, more pleasant work experience by transferring to a group that handles just emails or text chat encounters with customers. If telephone anxiety is what's hurting the work-life of the rep with great knowledge skills, take away his telephone.

12.Bring reps into the discussion when you are planning issues that affect them. Is there any reason in the world why the decision on which head-sets to wear has to come down from on high?

13.Encourage reps to become experts in one or more areas that have to do with your company. For example, if a rep is answering customer service calls about your product, the more that agent knows about how that product works, and how it connects to a myriad of other products, the more valuable the rep is when he or she gets on the phone with a customer.

14.Gather as much front-end information about the customer as you possibly can. There are so many good reasons to do this that it is pointless to enumerate them all. Suffice it to say that if you gather enough information to handle much or even most calls automatically, you'll be sending calls to the agent that truly require human intervention. When a person is asked to do something meaningful with his or her brain to solve someone else's problem, and in the process makes a person happy and improves the relationship that customer has with the company, that's what I call "good work." It's the kind of work that keeps a person engaged and therefore happy. This, in contrast to the state of mind of someone who is forced to answer the same repetitive questions time after time, day after day.

15. Train reps not just in the art of articulate customer service. Train them in how the center works. That is, explain to them exactly what's involved in call routing, screen pop, and how that call was chosen to arrive at their desktop. The more they know of queuing, hold times, and how decisions are made, the more reasonable seem the demands you make on them. Of course, some may see the downside here: if you do this you are going to have a lot of call center managers in training running around. And of course, if they know all this, they are certainly not going to be satisfied making $10 an hour.

Realizing the Value
On the Front Line

Let's start from the proposition that customer service in America is at best unremarkable, and at worst, really awful. Whether we're talking about service delivered by phone, fax, email, live text chat or website, it doesn't really matter. The typical consumer has a story to tell — often many stories — about the dysfunctional customer service systems that they have to deal with on a daily basis.

Now, before I talk about why this is or what can be done about it, let's have a moment of self-examination. It's my belief that the call center industry has a collective blind spot. We tend to screen out information about the interaction of the "typical customer." Instead, the industry as a whole, particularly people working at vendors of call center systems, often idealize customer interactions, or see things as if they were plotted on a graph or in a spreadsheet. Let me strip the veil of willful ignorance away from this picture for a moment: people hate this. We already know they hate being called by telemarketers; that goes without saying.

But they hate having to call an 800 number and wait on hold. They hate having to punch in an account number, and then have the agent ask for the account number again. (This is what CTI and its tech ilk have wrought — unreasonable expectations.) We too often view the industry, and the call centers that are its core, as a finished product. It's as if, having conceived of the best possible service situation, and enumerated the technologies that will get us to that point, we can sit back and rest on our laurels and marvel at what a wonderful technological age we live in.

It's not enough to simply say that now customers have choices about what channel to use, or to feel good because we can measure the value of customers and supposedly treat them according to how important they are to us. We must remember what I believe is a permanent, essential variation on the 80/20 rule: 20% of call centers (or contact centers, or whatever you want to call them) make good use of advanced technology; 80% do not. And another, more chilling variant: 20% of customers understand how the call center industry helps them, how the tech they wade through makes their interactions better; 80% don't care one whit, and hate the whole experience. (Note: these percentages are for illustration purposes; you won't find this in any study or research report.)

It's that 80% in both cases that come into contact most frequently, and when they do, watch out.

Let's also presuppose some basics. (Anyone who wants to dispute the following premises, please do so.)

· Most call centers do nothing but answer voice calls. No emails, no Web, no ecommerce. Nothing but plain old voice.

· Most first line agents are looking at a screen that tells them little more than the name of the person calling, and the current balance (if any) on an open account.

· Most of those agents are getting that pittance of information by manually entering a name or account number.

· Most of those agents are getting paid under $12 an hour.

· Most of those agents will be gone in three years.

· Most of those agents don't, then, have more than a cursory familiarity with the company they work for, or the products they sell or service.

I could go on, but why? It's clear to me that the industry, which for decades has been infatuated with technology, needs to squarely address the main barriers to continued success: a) that advanced technology is not so widely distributed as the industry press and vendors would have you believe, and b) the weakest link has been and remains the way reps are managed.

Treating the rep like crap is functionally equivalent to treating the customer like crap. To knowingly do so for very long is corporate insanity.

Fixing This Problem

One of the reasons CRM became so popular in the call center mind was because it contained the germ of a very good idea at its core. That is, you can measure the value of a customer, estimating the amount of revenue that a customer will generate over his or her lifetime. Then, when you balance that figure against either the ongoing cost of retaining the customer, or the cost of replacing him should he leave, you know exactly how much you can spend on customer service. If you have good knowledge of your costs and control over your operation, you can arrange to deliver service with pinpoint accuracy to the customer, always spending enough to keep him, but never more than you need. This is one aspect of what has been called the *mass customization of service*, and is a component of the popular idea of *one-to-one marketing*.

Well, it's all well and good to think this way. These are admirable goals, and we are at the very beginning of the era in which this type of marketing will become standard. We are not there yet, not by a long while.

Technology promises more than it delivers. I'm not talking about the specific features and benefits that are detailed in RFPs and planning documents. I have confidence that most vendors' products actually do what they say they do, in controlled or highly maintained environments.

But I don't believe that layering tech on top of a customer base will do anything but slow confusion most of the time. The call center blind spot tends to look at customers as aggregates, which is like looking at them as a herd. I don't think they tend to act as herds. I think that they tend to act as accumulations of a lot of highly eccentric individual actions. Each action is created by a counter-action, which is the conversation with the agent.

And that's why I propose that before you invest dollars and man-hours in a customer service system based on the value of customers, you attempt to measure, and understand, the *lifetime value of an agent.*

It ought to be possible, from the moment of hire, to estimate going forward the number of calls, minutes, interactions or customers that a given agent will deal with. Any call center that knows its own turnover rates knows how long an agent will stay for, on average. Correlate that with the ACD statistics for average call handling. Put a number to the agent's interactions.

Say you hire someone and you can predict that, based on turnover, the average rep stays for three years. Then say that the person will handle (for argument's sake) five calls per hour, or 40 calls in an eight-hour shift. That's 200 per week, times 50 weeks is 10,000 in a year, or 30,000 in a typical "lifetime." Now ponder that for a minute. How scared are you that the person you barely know is going to talk to 30,000 of your customers?

Thirty thousand of your customers, at the moment when the relationship with them is at its most vulnerable?

Forget CRM — customer relationship management isn't going to do anything for you, if you put someone in front of those 30,000 who thinks you couldn't care less what happens in his career.

Now put dollars to those 30,000 interactions. Depending on what you sell, that's either a lot of money saved, or a lot of opportunity squandered. Can the rep on the phone make the most of that opportunity by soothing someone who might bolt to a competitor? Or sell them something that they might not have thought of? *Do you even know what they are capable of?* If not, then you're looking at the question of value through the wrong end of the telescope.

So what does this have to do with the sorry state of customer service in America? Everything. Consumers demand more, and companies pay lip service to that demand by adding more technology that purports to decrease cost and increase the amount of contact a customer has — but this endless fascination with multiple channels of contact is a fool's errand. Make the calculation of the value of an agent — and I encourage people to actually create a metric for this, if there's none already — and you will feel differently about the dollars invested in recruiting and training.

This is almost a heresy, but start to think about agents in the same way you think about technology — as a system of customer contact. When you know what the agent's lifetime value is, you can think about the costs involved and then ROI them! More training? Better benefits? Strategies to reduce turnover and increase productivity? It all looks different when you think of the agent as the key component in the service equation.

The agent is not really the weak link in the process, so much as the thinking about agents is the weak link in managerial planning. Here's a tip: the best technology you can add to your center is not related to call handling, or Internet-enabling, or CRM. It's tech that enhances training, that adds coaching, that measures skills and schedules reps fairly.

Call center operations are increasingly accountable to higher management for their performance, from both a cost and revenue standpoints. Creating, and using, a metric for agent lifetime value is a great way to justify the headcount you need to perform. And it's a lot easier to realize the benefits of *agent relationship management* than CRM.

Part Seven

Outside the Center

It's hard enough to ensure

that your own employees

are doing the job right;

how do you make sure

that someone from

outside is?

Outsourcing

It's never easy to hire an outsourcer. Especially when what you are asking the outsourcer to handle is that most precious asset, the customer relationship. It's hard enough to ensure that your own employees are doing the job right; how do you make sure that someone from outside is?

Apparently, that fear is more widespread than I thought. A recent study by Input, a market research firm in Mountain View, California, found a disparity in the call center market. Users who were surveyed about their centers reported higher satisfaction levels with in-house call centers than with outsourced call center services.

Call centers will continue to grow at a strong pace but that growth does not come without a price. According to one study, users are worried about things like "staff competence," "flexibility" and "the caliber of operations" at their outsourced centers.

Before accepting that at face value, let's remember that the outsourcing sector is a very large component of a huge industry. It's also an industry undergoing a lot of change. Is their service awful? Not as a rule. Could it stand improvement? Without question. What I really think is happening is that the managers who decide they need to go outside their organization for help with teleservices are realizing how critical that customer relationship is, and are understandably nervous about losing control over it.

There's no question that the mere act of turning sensitive service and revenue tasks over to an outside vendor creates stresses that boomerang back as tentative satisfaction ratings. "As an industry, in terms of keeping our own house in order, there's probably some truth to that," says David Kissell of Alert Communications. "But the client can't just hand off the account without paying attention," he says.

The fact is that the outsourcing business has boomed in this decade. Companies need the service, the expertise and the technology that outsourcers make available.

For the outsourcers themselves, this has been a strange time — people are beginning to realize what a fragile business model call center outsourcing is. Outsourcers are at the mercy of a customer base that demands the highest levels of technology, that insists the outsourcers provide incredibly sophisticated off-premise technology, and that they integrate it into existing systems.

On top of that, outsourcers perform a service that, to most of their clients, is a luxury — that during bad times will be scaled back to cut costs, and that (I think) will prove to be very price competitive. Every negative thing that affects call centers hits outsourced centers harder: a shortage of qualified labor, the capital costs of keeping up with demand and new technology, and the capricious introduction of unproven innovations (like Web/call center combinations) that the market calls necessary but that are a killer to do properly. Is it any wonder that the few outsourcers that are public, and hence report earnings, have a devil of a time making money?

Sometimes, when a company is growing, the only way to keep up with an expanding customer base is by getting help from outside. Traditionally outsourcers served as a bridge — handling high call volumes during peak seasons, for example, or during product launches. Additionally, service bureaus are a good way for a company to test something new without incurring capital expenses. A new campaign can be tested on an outbound list without buying dialing equipment, or hiring new workers. Outsourcers, to stay competitive, often feature the latest technologies in the most sophisticated implementations.

But just as the relationship between a company and its customers needs constant managing, so does the tie between a company and its outsourcer. "What we have to do is when someone brings us an application, go through it from square one to identify their goals and objectives," Kissell says. "The commodity mentality of the [service bureau] product has changed things," he says.

Instead of treating every client as if they were punched out by a cookie cutter, service bureaus have to invest in a lot more handholding with their clients. Outsourcers' problems tend to get magnified by the attention those companies get in the press, and by the fact that in recent years several high profile companies went public. In general, service bureaus run centers that are larger than those run by companies in house. They also tend to run more of them, often connected into networks of interlinked centers. They tend to be as afflicted by things like high turnover and employee burnout as any other sector of the teleservices industry.

There are some things a company can do to make sure they get the most out of their relationship with an outsourcer.

First, they need a clear view of what they expect the outsourcer to accomplish. "They need to find an outsourcer that is a partner, and not just paying them lip service, who makes a concerted effort to understand what their goals are and what's going to make their program a success," according to Kissell.

He points out that a lot of service bureaus make the mistake of pigeonholing clients; that is, assuming that a single basket of services will suffice for all catalogue retailers, another for hospitality companies, and so on. "A lot of service bureaus will fail this way," he says.

End-users should pay close attention to the specialty of the outsourcer they're considering. Are they experienced doing the kinds of things that you do, and if so, are they coming into the relationship with preconceived notions of how you should be running your business?

Close attention includes checking references and calling into centers to see how calls are actually handled. Ask the outsourcer about staff training — is there a regular program for refreshing the knowledge of phone reps? And how about turnover?

They should be able to tell you what physical centers will be used for your campaigns, and what the turnover rate is at those centers. You have a right (and a duty) to find out how skilled, and how motivated, those reps are. Ask what kind of career path is available for agents — do they get promoted to supervisor? How long is the average tenure? Knowing the answer to those questions will tell you how bored the rep is when he answers the phone and it's your irate customer on the line.

I suspect that the disjunction between satisfaction at in-house versus out-sourced call centers is due more to "location comfort" than anything else. You know the people who work in your own call center, their strengths and weaknesses. You've probably already figured out how to play to those strengths, deal with the weaknesses. Admitting that you need to go outside for help, especially for the smaller, growing company, can be dispiriting. Losing touch with the process of customer contact, even if it's limited, can be unnerving.

Service bureaus need to emphasize the connection they offer between the customer and the company. I think they have a perception problem, not a real problem. They need to sell their experience, their consultation. There are technical tools at their disposal that enable the companies that hire them to come closer to the point of interaction — real-time reporting tools, for example, that let a client see the results of calls (inbound or outbound) without having to take those calls themselves. There are monitoring and quality assurance systems that deliver voice and data records to the client, if needed, of everything that happens during a call. Outsourcers can be — and should be (and I think are willing to be) held accountable to their clients.

Most important, Kissell says, is that they talk through the relationship before it even starts. "There should be handholding before the ink hits the paper," he says.

■ Is Your Telco Your Next Outsourcer?

What if you ran a call center and your network carrier offered you the same services — call handling, transaction processing, and order fulfillment? What if they offered to consult with you on creating the most efficient center, and using the center to support your company's strategic goals? Or offered you the chance to offload some or all of your volume to their centers?

What if you could take it a step further, and use the carrier network itself? After all, you may have an ACD, but they have better ones. That's what the phone networks are made of. What if all the queuing, routing and call processing — in short, every touch of the customer from the first IVR interaction to the faxing back of an order confirmation — what if everything could be handled by AT&T or Sprint or even British Telecom? Wouldn't that be at least as attractive a proposition as going through a traditional outsourcer?

Carriers have a powerful incentive to get into this business. First, the expertise is already there. They know how to handle calls and call centers; some of their centers are among the world's busiest. The long distance carriers have long used their own centers as test beds for their own new technologies, including some of the enhanced network services that make their entry into the outsourcing field possible.

Second, the economics are compelling. Carriers thrive by selling telecom minutes to call centers (to others, too, of course — but 800 number traffic, the bread and butter of call centers, is also a key part of their revenue). Anything they can do to generate more usage of their networks is money in the bank. If they offer a call center an off-premise solution for IVR, for example, or a multi-site option that lets the company hold calls in the network while waiting for an agent to become available, that generates minutes. Maybe they'll craft you a discount that brings your long distance costs closer to zero cents per minute; they'll be able to do this because you're paying for the value-add, the service that happens in their network.

This works best when carriers offer software-based services that don't require them to spend on agents. But over time, the carriers will evolve their call center offerings into as close to turnkey call processing as anyone has ever gotten. Already I've seen carriers get into the act as call center "consolidators" — fitting together all the technology pieces under one umbrella purchase. Come here, the sell goes, and you can get up and running quickly with a center that we will help you build, and you can pick from a Chinese menu of hardware and software vendors to supply the applications you want to run; we (the carrier) will certify that everything integrates happily, that everything works together and you can call one number for multi-vendor technical support.

But it's clear that the role of the carrier is going to take on more of what was traditionally expected from an outsourcer. The carriers can craft these offerings, and push other vendors into working relationships, because 1) they are large; 2) they can set standards; and 3) they have a relationship with the call center that runs deeper than that of the traditional outsourcer.

Of course traditional outsourcers are not going to go away. The fact that carriers are superbly positioned to get into this business does not mean that they will do so in any meaningful way. The fact that they have had the means and the opportunity for six or seven years now and we are only just beginning to have this discussion in the industry means that the competitive advantage still lies with the traditional outsourcer. Carriers are only now awakening to the possibilities of enhanced services. But they are notoriously bad at crafting products from technologies.

Carriers can be slow, cumbersome, anything but fleet of foot. They can be danced around by clever competitors.

From the call center consumer's point of view, the more choices the better. All the automated front-end transactions, especially IVR and routing, can be well handled outside the call center. I think companies will look to outsourcing as a specialty: if an application has more to do with routing and automated call handling, the carrier will be a better choice. But if it's more agent-based, and involves touchy things like selling or servicing existing customers, a traditional outsourcer will probably get the call.

It's virtually certain that we are in for a few years of amalgamation, where the lines between different types of outsourcing blur. Who will offer Internet-based transactions, for example, or video-enabled call centers? What's certain is that they will all compete to offer a wide variety of new and improved services, and call centers will be better off.

Ovum, an industry research group, recently reported that the biggest growth segment in the call center market was for network-based call center services. They say that these services will generate more than $4 billion in annual revenues for telecom service providers by 2005. (By contrast, current worldwide consumption of toll-free services by call centers is put at $20 billion per year.)

On top of that, they say, 35% of call center agents worldwide will use some type of network-based call center service, with nearly half of those using network services as their primary call distribution method.

Network-based services refers to any agent-support systems that traditionally occur within the center: things like call routing, transaction processing, database lookup, screen pop, and so forth.

The interesting thing about all of these things is that they can be — and are being — done by the network carrier in the telecom network, outside of traditional premise-based call center equipment. Carriers see this as a potential gold mine; they can drop the price of toll-free to just a hair's breadth above zero, and more than make up the difference selling you services as a value-add.

The kinds of things that can be done in the network include what I think of as virtual or distributed call centering — dispersing agents among many centers and routing calls among them as if they were co-located in one site. It also includes network IVR, grabbing the customer input in the network, then using that to determine what to do with the call.

Maybe most important, Web integration services hold a lot of potential for this. Call centers are having trouble coping with both the technical and human resources issues that are cropping up due to the explosion of Web access channels into the center. Adding things like live text chat, call-me buttons and even simple email create headaches for reps and managers alike.

Putting some of these things in the network (particularly the IVR) is something the telcos themselves have been doing for years. Add Centrex ACD and the hunger for multi-site centers and you have a tempting service market. From the call center's point of view, network-based services that are paid for either by the month or by transaction offer a way to be more flexible in the face of unpredictable volume and varied access pathways.

In the end, I think this will be an opportunity for call centers to play mix-and-match with their technology and outsourced services. I think network-based services will offer an attractive alternative to premise equipment for a lot of centers, and open the door to new ways of managing centers.

■ What to Look For

We used to think of outsourcers as predominantly outbound entities, doing mainly specialty telemarketing. That's not the case anymore. In fact, the outsourcing functions of a service bureau now go far beyond what we traditionally expect. Everything we think of as "back office" is now fair game for an outsourcer: everything from inbound and outbound call handling to customer tracking, quality assurance, fulfillment, data processing and even help desk customer support.

Or should I say: especially customer support. More and more, companies are turning over their help desks to outside experts. It's easier than it's ever been, thanks to technology that assists in routing and tracking calls. And it's more cost-effective.

As post-sales customer support becomes simultaneously more important and more expensive, companies are looking for lower cost alternatives that don't force them to compromise on quality.

What do you get with a service bureau? A few deceptively simple benefits:

· Access to technology. Forget about the big capital investments in switches, dialers, workstations. In upgrades to hardware and software. Service bureaus are equipped with state-of-the-art call center systems. They can spread the costs around multiple clients. Unless you devote strong (and consistent) resources to your in-house call center, only a service bureau will have the cutting edge technology you need to stay competitive — and have it in a hurry.

· Vertical expertise. These companies are specialists. In the needs of banks. Or fundraisers. Or retailers. Whatever industry you are in, there are outsourcers who know how that vertical market functions and how to treat your customers.

· Speed. You can respond quickly to seasonal (or even hourly) fluctuations in the number of agents your program will need.

One of the surprising developments in recent years is the way certain high tech companies are growing into major outsourcers themselves. This is due to the need to provide intensely consultative service to many of their own customers.

As call centers change (becoming more distributed and taking on more business functions) the companies that provide them with services change too. There is much more emphasis on advanced computer telephony integration technology at outsourcers these days. And on technology for linking call centers with other back office processes. Outsourcers are now, more than ever, the engine of growth in the call center industry.

Outsourcing service bureaus face many of the same pressures as in-house call centers. Outsourcers are a good gauge of the cutting edge. They've always been in the forefront of technological and operational change in the call center industry.

In fact, even before there was such a thing as a call center industry, service bureaus were busy creating it — carving out an identity for call centers that had less to do with the vertical market served than it did with the techniques involved in handling calls.

What will service bureaus look like in five years? Not terribly different, at first glance. While there are several trends pushing the call center in virtualized, dispersed directions, the physical reality of today's centers — rooms full of people, talking into headsets — won't change dramatically.

The changes in service bureaus over the next few years mirror what's going on in the rest of the call center industry. They all face the pressure of improving productivity and delivering more services directly to the end-user. And they are all scrambling to provide more "self-serve" methods of interaction — letting the customer search a database for the answers to his own problems, for example, or use an automated system to transfer funds, and most especially, they are rushing like demons to offer Internet front-end services, and to integrate those services with back-end database tools.

Available to all these centers will be an evolving basket of powerful technologies, some new, some enhanced. The difference for outsourcers is that they have even more pressure than the rest of the industry to stay ahead — to use those technologies to eke out even tiny efficiencies. To make money in that small spread between what it costs to handle calls, and what you can charge the client for that service.

■ Specialty Niches

One other important reason to go with an outsourcer is to crack a specialty niche for the first time. Are you prepared to market to the Hispanic market with your current staff and configuration? How about Europe? Or even Canada? There are service bureaus that specialize in Spanish-language tele-marketing or other multilingual options.

Others take on the needs of a particular industry, like fundraising, collections or high-technology.

Disaster &
Contingency Planning

Fire, flood, work stoppage, electrical uncertainty — there are acres of reasons to worry about the ability of your center's critical systems to function non-stop. What you need to do is plan ahead for as many contingencies as you can think of. It's not a luxury — it's a necessity.

There is no such thing as "good" downtime. In a call center, downtime means calls aren't coming, orders aren't being taken, customers are getting angry, impatient, upset. Or they're simply calling someone else, and you'll never know about it.

There are a lot of things that can bring your call center to a screeching halt. (Actually, the halt will be very, very quiet. . .) Natural disaster can keep people from coming to work — storms, snow, flooding. A construction crew digging two towns away can cut your long distance link to the outside world.

Or, in perhaps the most common situation, problems with the electrical power system can knock you off line in one form or another without warning.

Here are some of the ways that centers (and vendors who supply those centers with technology) can protect themselves.

1. Identify the key systems that are at risk. Most are obvious: the switching technology, the data processing. But do you know how vulnerable your business is if your package delivery service isn't available? A lot of call centers found out the hard way when UPS was hit by labor trouble several years ago. It's one thing to be able to take orders by phone, but when you can't fulfill those orders, customers may turn away.

   External services that are outsourced are particularly vulnerable. For better or worse, companies are bound to other companies in dependent relationships for critical applications like fulfillment, personnel supply, even for service on internal equipment.

   If you rely on outside services that you deem critical to continued operation, you have two choices. You can single-source that service and make certain that the vendor has enough redundancy (and extra capacity) to handle a problem. Or, you can multi-source, giving yourself a backup in case your main vendor has problems.

   This applies to everything from package delivery to long distance services — if you rely on AT&T to bring calls into your center, and a backhoe in New Jersey cuts all AT&T traffic, it's nice to know that some of your customers can get through by using Sprint.

   And if you are a service provider, let your customers know what contingency plans you have in place to assure them of continued service in case of snow or fire or other short and medium term emergency.

2. Do a cabling/wiring/power assessment. I suggest you map out every wire and connection in your center — and diagram the connections between technologies.

   What this will do for you is, first, allow you to check the power protection status of every server, PC, switch and node in house. A critical assessment will tell you which are covered by UPS units, which have hot-swappable power supplies, and which need them. Do you have phone sets that get plugged into bare outlets? Guess I don't need to tell you how awful it's going to be when lightning shorts out all the phone sets and headsets, leaving reps with working computers and incoming ACD calls they can't answer.

That kind of thing can happen to the average call center on any day of the week, any time of year.

3. Identify key personnel. It's important to know who will be on call during a problem situation, and what those people's specific responsibilities will be. It's even more important that they know what they are supposed to do and where they should be.

It's going to be demoralizing to go around the center and the organization looking for points of failure. But once you do this, you have to marry the possible problems with the people who can do something about them.

Any working group you convene for call center contingency planning should include members from outside the call center, especially including people from IT and facilities management departments. Share knowledge. They need to be made aware of the impact call center failure could have on the entire company. (Often, amazingly, companies are blind to the fact that a call center outage will affect revenue.)

And those people outside the center need to provide coordinated responses to problems that affect data processing, order processing, shipping, human resources availability — the whole gamut of business functions that are involved when you're trying to jump start an operation as complex as a call center.

4. Identify manual workarounds. If the computers went down, could you take orders by hand? Or would you have to tell customers to call back later? Make sure there are enough copies of your current catalog or product list for every inbound agent at all times — if they can't pull up product info on the screen, at least they can give a caller basic pricing and ordering information.

Do you have a plan for coping without a CTI link? Agents should be trained on the procedures for handling customers when they don't have access to customer data.

What about when the front-end crashes — how will you handle the sudden flood of calls that come into the agent pool when the normally reliable IVR or auto attendant doesn't siphon off half your calls?

Or the Web? How will your center react to an inflation (or deflation) of the contact volume through alternate means, like email, or text-chat? The more access methods you provide for customers, the more points at which a sudden change can cause problems — not true disasters, perhaps, but certainly unintended difficulties.

On the other hand, the more avenues a customer has to get through to you, the less likely you are to lose that customer to a disaster. You can cross your fingers and hope that if all your call centers shut down for three days, they'll still find you on the Web.

5. Explore secondary sites. Sometimes, the only way to truly prevent disaster is to replicate your call center functions somewhere else. If there's a flood in your center but your people can still come to work, you can continue to operate from another location.

This could be as simple as using non-call center assets (basic office space, for example), or as complex as arranging to buy contingency services from someone like Comdisco.

In fact, you can arrange (from a number of reliable vendors) to have the entire call center run from alternative sites in any location, for a hefty fee. These "call centers on call" are different from traditional outsourcing, in that you pay a retainer to have access to the services on your schedule, but you only call on them when you really need them.

These disaster-oriented services can provide equipment, temporary (and sometimes mobile) facilities, and if you're inclined, data processing and backup functions as well.

What's most important to remember about any of the options described here (and I've only just scratched the surface) is that the continued operation of your center depends on a complex set of connected technologies, some of them very fragile. Some are vulnerable to things you can't control.

What you absolutely must do is take some precautions to assure a minimum continuity of function and connection to your customers. How you do it will depend on your unique circumstances, but that you should do it is unquestionable.

The results of any of those events could have devastating effects on a company. The call center is one of your company's most vulnerable segments because it is the nexus of several complex technologies. Losing it means you are cut off from your customers — so in any disaster it should have top priority for recovery.

◼ Power

Power protection is probably the most serious ongoing problem. When a call center goes down, revenue stops coming in. Protecting yourself from disasters is one of the least expensive — but most neglected — aspects of call center management.

Your IVR system is a perfect example. Depending on how it's implemented, IVR can quickly become a crucial part of your enterprise — handling a substantial amount of your call traffic, promoting customer satisfaction, generating revenue. In some cases the IVR system actually touches all inbound calls, either handling them outright or passing them back to the agents in conjunction with an ACD. In this context, IVR downtime would be as damaging to operations as a loss of phone service.

Some things you should know:

· Power problems are the single most frequent cause of phone and computer system failure. The average IVR system, for example, is hit with a significant power fluctuation (spike, surge, brownout) approximately 400 times a year. The problem is getting worse, as regional power grids are forced to adapt to increased consumption.

· Power-related damage is among the most difficult types of damage to recover from. That's because it does two things: it cripples the hardware, often necessitating a costly replacement (complete with waiting time for delivery), and it wipes out data.

· Multiple connections (to trunks, networks, peripherals, etc.) increase the number of routes through which power surges can enter and cripple an integrated call center. The more components you connect together (and I count data sources here as a component) the more vulnerable you'll be when something hits.

An Uninterruptible Power Supply (UPS) is a battery system that provides power to your telephone switch or computer. Surge protectors or arresters, clamp down on high voltages that can surge down your power or telephone line, frying your delicate equipment. Power conditioners remove noise, make up for undervoltages, and generally deliver clean power to your telephone switch or computer.

Many high-end UPS systems roll all of these functions into one unit. The big news in power protection these days is power management software, which lets you keep track of power conditions throughout your network from your workstation, and gives the UPS more sophisticated features, such as the ability to shut down your equipment even when it is unattended.

Many power protection systems (especially UPS systems) are marketed specifically for telecommunications applications, but the fact is, the same UPS can protect a telephone system as well as a computer system.

Power protection is one of the cheapest forms of insurance you can buy. The technology is solid. It's been proven to work. The added cost of protecting hardware is roughly 10% to 25% of the hardware's value. That does not count the value of the data. You could make the argument (I often do) that a call center's data is far more valuable than its hardware.

When you factor in the cost of potential losses, and weigh that against the likelihood of problems, the cost of protection becomes negligible.

■ The Audit

There are two parts to a successful disaster strategy:

1. **Contingency planning.** Making sure that you've identified all the people, processes and equipment that are critical to your operation. Plan for their unavailability, with backup strategies for a variety of situations.

2. Installing a technology net. Power protection. Backup power supplies. Redundant trunks and carriers, etc.

Here's what I think you should do to protect yourself.

· Document everything. From where your wiring runs to the home phone numbers of all critical personnel. Put your plans down on paper — yes, paper — so it survives a network crash. Rehearse them. Tell everyone where they are. Think about how stupid you'll feel if you go to the trouble to create a plan, then can't find the information in a pinch.

· Identify risks. It's not the same for everyone. Geographically, you could be more likely to encounter a snowstorm than an earthquake.

Also, it's important to identify exactly what you need to protect. In every business, there are systems that are supercritical, and those that you can live without for an hour (or a day). Establish your own priorities — is it more important to be able to take orders? Or provide service? Thinking about these priorities will also give you useful insight into the way your call center fits into the company's business process.

· Get a power audit. UPSs (uninterruptible power supplies) are designed with the assumption that building wiring provides proper routes to ground and has sufficient ampacity to sink diverted power surges.

But the best UPS in the world won't protect you if your building wiring is crummy. Neither will the "guaranteed" insurance policies that come with some power protection products (read the fine print).

Find an objective electrician to conduct a true power audit of your premise wiring, and make necessary repairs and upgrades part of your power protection program.

· Protect everything. The object is to set up an island of protection around each crucial piece of equipment. That means UPS protection behind the system. And properly grounded surge suppressors on network and phone connections, as well as cables and connections to peripherals.

Other department managers can get away with focusing on one or two critical elements, but your disaster recovery plan must cover the protection and restoration of:

· Telecom equipment (hardware and software)

· Computer equipment (hardware and software)

· Networks (voice and data)

· Electric power.

And that's just the technology. A complete plan will also include contingencies for your facilities and your personnel.

■ Telecom

Assume your phone system will fail, and at the most inopportune time. I strongly urge you to get involved with your local phone companies and the network carriers to create recovery plans. Most outages occur on the local level.

Also, make arrangements with your equipment vendors to have emergency replacement equipment available on short notice. There are switch vendors who will take a machine off the assembly line for a customer that's close to what's in the field and build it quickly into a machine that's suitable to the loss.

Have lots of backup power supplies in place, because most emergencies are due to disruptions of the power grid. That includes batteries, generators and UPS systems.

There are four specific areas in which you need to be prepared:

· Have a plan documented.

· Maintain the call center's files and knowledge bases backed up, current and off-site.

· Have a place to go to reestablish the call center, properly equipped with computers and telecom equipment.

· And critical: train your people.

Switch manufacturers build in disaster recovery features through system redundancy, service options and fast turnaround on replacement systems.

The favored way to protect your networks, especially your telecommunications networks, is through route diversity. Arrangements for network diversity are made with your local telephone company, an alternative service and your long distance carrier.

At the local level, you don't want all your traffic to be carried over the local loop supplied by the local telephone company. The local loop is vulnerable, as is any network connection, to breaks because of construction (backhoe through the cable), destruction (a car crashing in to a telephone pole) and other disasters.

Your local carrier can usually supply an additional loop. Make sure route diversity extends all the way to the point where the trunk enters your premises. Real routing diversity involves at least two entry points into your buildings.

Some companies achieve diversity in their long distance service by using two or more long distance carriers. And it's a tactic your carrier may not tell you about. Long distance carriers have their own tactics for network diversity that are worth asking about.

Of course, third party suppliers are available also. All of Comdisco's recovery facilities are linked through a high-speed fiber backbone network called CDRS Net. This network is also available to clients to recover voice, data and image communication links between the client's site and Comdisco's facilities.

Another point of concern is the interface between the telephone network and your equipment. Not much of an issue on your home telephone, but in a call center with several T-1s, you can't forget about protection for all the sophisticated equipment that connects the T-1s with your telecom equipment.

Protecting a telephone system from power disturbances that originate on the AC electric line is common practice at many companies. What is often forgotten, though, is that your telephone lines are also a power source that can introduce voltage surges into your system or be blacked out by lack of power.

With regular telephone service, the problem is that your telephone switch has no power even though the telephone service itself is working. There are protective devices on the market that let you bypass your telephone system during a power outage by connecting preassigned single-line station phones directly to the central office trunks.

This won't be much help in a large call center, but is just the thing to let a small call center with only a few lines keep handling calls through a power outage.

Telecommuting Agents

Faced with high turnover rates and a tight labor market for call center labor, some in the call center industry are renewing an idea that seemed all-but-dead just a few years ago: the notion of call center reps working from home.

It didn't die because of feasibility. In fact, for most of the last ten years it's been possible to wire up call center reps to the center from just about anywhere, first through ISDN, then thanks to the Internet.

Several major ACDs were open to that kind of link, and the software to remotely manage the agents could be had, albeit for a price.

No, what pushed it aside was the fact that it removed agents from the direct control of their supervisors, and often it was difficult to show quantifiable reasons why you should go out of your way to do that. In any industry, letting people work from home is a giant step for managerial culture. In call centers, it's the equivalent of an earthquake.

And yet, subtly the idea is catching on in the rest of society. One study showed that in 1997 there were 11 million US telecommuters, up substantially from just a few years before.

That same study (by the International Telework Association) suggests that telecommuting's feasibility is enhanced by access to advanced technologies, such as the Internet and personal computers. An estimated 31% of telecommuters used the Internet, more than double the average home usage rate, and 75% of them used personal computers, up from 59% in 1995.

Two things to take away from that stat. First, it was for 1997, which as we know, was only the beginning of the Internet wave. We would expect those percentages to be higher now.

Second, that at the same time the Internet is reshaping the nature of "work," it is reshaping the organization of the call center. As these trends dovetail, the call center becomes a more attractive place to implement creative job structures — if you can wean managers from having daily physical contact with their reps.

There are powerful arguments in favor of having some portion of your call center agents telecommute. One of the biggest is the morale boost — and the resulting money you'll save on training when they stick around for a long, long time.

Call center workers are ideal candidates for telecommuting. Anecdotal evidence suggests that most people who work from home do so because they are employed by small companies with less than 100 employees.

One reason why smaller companies lead the way is their informal management style makes testing the idea easier. The companies that benefit most from having employees work from home are those which have a workforce made up primarily of information workers or service workers. Many call centers have employees that fall into both categories.

Switches (and their software) have evolved to the point where agents can log into an ACD and receive calls at home in exactly the same fashion as if they were sitting at a desk in the center. Their desktop can receive the same screen pop of data. And most important, the agent appears on the supervisor terminal in real-time — so they can be counted, evaluated, monitored, and communicated with. Switches from Aspect and Rockwell, among several others, offer advanced and transparent home-agent capabilities.

■ Why Consider It?

Why would anyone spend even a penny to have agents work at home? There are several good reasons.

1. Gain productivity. Trials have shown that workers are more productive at home. The main reason for this may be that there are fewer interruptions. The fact that the telecommuters are more comfortable at home and are avoiding the stress of commuting may also contribute.

2. Reduce training costs. There are many advantages to decreasing turnover, but the bottom line benefit is the reduction of training costs. One company saved over $10,000 per telecommuting employee. The bulk of that was money saved on training.

3. Retain employees when they move. With two-career couples now the rule, not the exception, it is easy to lose a valued employee because a spouse's job requires relocation. Companies can retain the knowledge and experience of these workers by having them telecommute from their new home. (Obviously there are some limits — you don't want to create a continent-wide network of agents, that's just too unwieldy.)

4. Retain employees with family obligations. Today's workers have obligations to both ends of the age spectrum. A worker may leave a job to care for a child or elderly parent.

   While working at home is usually not possible when young children or seriously ill adults must be cared for, flexible work schedules let telecommuters fit in part-time work when they are free from their other responsibilities.

   This also allows you to tap a hidden source for call center staff — the spouses of your reps. They know your business, and may only need minimal training before they are ready to serve as part-timers on-call in a pinch.

5. Provide call coverage in emergencies. Earthquakes, snowstorms, floods. Almost every region has experienced a natural disaster that made it difficult or impossible for agents to reach the call center. Agents can work at home temporarily when disasters disrupt roads or damage call centers. It provides you with an instantly accessible backup plan for many low-grade problems.

And because the overhead costs are so low, temporary at-home agents are also a cost effective way to deal with peak periods — whether they are expected or unexpected.

6. Provide a new source of employees. The largest call centers often employ all qualified agents within a reasonable commuting distance. But what happens when those call centers need to expand?

Adding at-home agents that are outside of traditional commuting distance is one way of expanding the workforce without the expense of opening another call center.

7. Comply with the Clean Air Act. This federal law requires companies with more than 100 employees in high-pollution areas to design and implement programs to reduce air pollution. Telecommuting is a sure way to reduce auto emissions.

8. Comply with the Americans with Disabilities Act. Another federal law gives disabled workers the right to employment in jobs they are qualified for. Agents that can't commute because of their disabilities can be accommodated through a work-at-home program. At-home employment can also boost the productivity of other disabled workers.

One thing most companies with successful telecommuting programs have in common is the ability to successfully manage their workers remotely. The main reason companies don't explore telecommuting is fear of lack of control. But with today's systems supervisors can monitor agents' work as if they were sitting in the call center. It is possible to randomly access both computer screens and telephone calls as if the agent were in the same building.

■ Who Gets the Nod?

It is important for a company to put prospective telecommuters through a selection process. Agents should *earn* the right to work at home.

You should have stringent requirements for the selection of agents that participate in a work-at-home program. First, reps should be evaluated on work performance, including factors like attendance, promptness and productivity. Then the agents' homes might be inspected for size, electrical wiring and other factors important for choosing an office site.

Experts all say that managing the expectations and feelings of those workers "left behind" in the office is a key part of any successful telecommuting program. It is not just managers who have to know that the work-at-homer is producing. Co-workers must know also.

Obviously this is not something that every center should consider. It's not right for every agent. But I think it's a good idea to have at least some capacity to plug in agents from home, if only for redundancy and peak coverage.

Imagine: a sudden call spike comes in at some strange, unpredicted hour. You do not have the staff in the center, and it would take you at least several hours to get enough people in the seats to cover it. Wouldn't it be great to be able to call ten people at home and ask them to log in from their PCs? Pay them a great overtime rate, and you've solved a myriad of problems at one blow. Extra cash in their pockets, better service level provided, your stats look great, and it's all achievable with very little extra technology. Something to think about.

■ How To Make It Work

One thing that's going to make at-home agents a more viable option in the near future is the expansion of browser-based technology. More and more, the browser is the engine of choice for delivering data to the agent desktop. Though I've seen it used for quite some time for ACD stats in "readerboard-replacement" systems, it's being adopted as a way of getting actual call information, sometimes even including CTI pulls.

La Jolla, California-based Telephony@Work offers a browser-based, thin client interface for its *CallCenter@nywhere* platform. The browser-based agent interface is a multimedia version of the *CallCenter@nywhere Interaction Manager*. The application will run on any browser-equipped computer and therefore requires no pre-configuration, downloads or other executables.

This means that theoretically, you can locate people literally anywhere and they can still grab any of the data they need to perform their jobs. And more important, agent software is computer independent — meaning it is not tied to a specific model or operating system. This also saves time and money, because there are no configuration issues at the agent desktop.

Distribution and updates of the agent client software can be delivered and updated instantly.

This gives ASPs, carriers and campaign-heavy call centers a way to "go live" with little or no advance technology preparation. It's also a step toward the endless decentralization of authority in companies that run call centers; at this point call center management ceases to be an issue of managing telecom technology (where most call center managers cut their teeth in the 1990s), and instead becomes a combination of IT management and Human Resources.

CallCenter@nywhere is a comprehensive contact center platform for enterprises and ASPs that applies skills-based routing discipline to inbound phone calls, faxes, email, voice mail, chat sessions, Web call-back and voice over IP transactions with blended outbound predictive/preview/progressive dialing capabilities — in a single engine that allows individual agents to alternatively handle multiple mediums of interaction from a single blended seat.

The platform also provides Web-collaboration/form sharing and quality monitoring capabilities. Its got a neat feature: a hot-backup architecture that provides uninterrupted service for both customers and agents even if individual servers go down.

The system remotely communicates with the *CallCenter@nywhere* multimedia server from anywhere on the LAN, WAN or Internet. From inside the browser, agents see visual icons representing the different available types of media interaction — voice calls, voice-over-Web calls (though why you'd want to alert the agent to the difference seems silly), emails, voice mails, chat sessions, faxes and Web callbacks — and can control these interactions at will from anywhere. These "calls" can be answered, sent to voice mail, transferred to other agents and totally controlled as if the agents were co-located with the *CallCenter@nywhere* server.

And as you would expect, a browser-based client app should be able to handle calls that come from either packet switched or circuit-switched telephone networks. It supports both calls, acting as 1) an interaction control dashboard with built-in Voice-Over-IP soft phone; 2) a multimedia interaction control supporting analog agent extensions; and 3) a complete multimedia interaction control with synchronized control over telephone calls that are forwarded to separate (regular) phone lines — acting as a software-based PBX-extender that synchronizes call control and data flow over the Internet.

Calls are forwarded separately to home-based agents over the public switched telephone network (PSTN). This last configuration is important for contact centers wishing to deploy home-based agents with insufficient bandwidth to support toll quality Voice over IP communication. All configurations are supported in a blended environment, to support both on-premise and remote agents simultaneously — and packet switched Internet-based calls can even be conferenced with traditional PSTN calls.

It provides a real-time readerboard display of all multimedia queues for the workgroup associated with the agent. This includes information on the number of agents logged-on, agent availability and the "in-queue" counts for pending chats, phone calls, emails, voice mails and faxes. This stream of information never gets in the way of the actual call and media handling the agent does with customers. Each frame is separate and provides instant screen refreshes for each function.

Self-Service and Self-Delusion: A Final Word

Why is everyone so convinced that customers are delighted to have the opportunity to serve themselves? There are lots of ways to justify why they *might* like it, and why it's in the interest of any company to have customers that like it. But it's not so clear that they actually do — instead, it appears to us that the customer service industry has fallen into the trap of believing its own wishes, and then putting tools and practices into place in hopes that those wishes come true.

Here are the premises on which this possibly-flawed assumption are based:

1. Self-service is cheaper to provide than agented-service. True. Lots of evidence exists to support this contention. And it seems self-evident that a person who finds an answer through a Web interaction will not rack up the kind of costs associated with a long phone call. The consensus is that on a per-interaction basis, nothing costs as much as agented-phone service. Every other option, from email to Web to chat, costs less.

How much less is a matter of conjecture, though. Most alternatives to the phone also involve agents; it's just that those agents are handling data rather than calls, and are supposedly handling more of them per hour than phone reps. When you add the massive costs of the new technology infrastructure to the continuing ongoing cost of agents, our guess is that you don't come out that much better. All you've really done is add another channel of access, another layer of management over the new technologies (because you're never going to get rid of the phone component), and added back-end integration to your list of headaches. Worth it? Maybe. But a significant cost savings? Hardly.

Even so, most of these alternatives don't even qualify as "self-service." They are different, allegedly cheaper ways of reaching repped service — call them "Service Lite."

2. All these interactions are supposedly made more efficient by the fact that when a person finally does reach an agent, he has spent time gaining information from the automated systems, and therefore the call is shorter, more to the point, and ultimately better.

That's a great theory. Does it work in practice? No. Our anecdotal experience — and it has to be anecdotal because no major study has shined its light into this area — is that people who work their way through an automated system and then still have to talk to an agent are more irritated, not happier. If they are better informed, it's about how badly they are being treated, and how little the company really cares for their business.

The *New York Times* reported (in late 2000) on a study by the University of Michigan called the American Customer Satisfaction Index; since 1994 this measure of customer satisfaction has dropped in nearly every industry sector, and it's been blamed on the increased use of technology.

There you have the key disconnect between companies (and the vendors that supply them) on one side, and the customers on the other. It's just a pitiful lie to say that you are installing an IVR system in order to make people's interactions with you better. The truth is that you're doing it so you don't waste agent's time with repetitive stupid questions that cost you money to process. That's not a pretty truth, but it is a truth. Telling the customer that its for his own benefit is a piece of spin that few believe; the ones that don't believe it come to think of you as distant and uncaring.

3. Customer relationship management bridges the gap between company and customer by enhancing the agent's responses at the moment of interaction.

This is so not true people in the real world (i.e., customers) would laugh themselves silly if they understood what was really going on behind the curtain.

First, the real goal of CRM is to coordinate data behind the scenes. It's a way of getting different departments with different sets of data to push all their information into one big pile, and then analyze it so you can have more efficient customer interactions. Efficient: cheaper, shorter, and producing more revenue per customer.

It's also designed to provide a way to analyze the tracks of interactions that come in through different channels, so that a single customer can be seen wholly, no matter what he's doing with your company.

These are good goals, from the point of view of the organization. But what they are not about is making a better interaction from the customer's point of view. One of the chief uses of these tools, by one of the largest adopting sectors (financial services) is to create scores that rank customers based on their value to the company. And when you do that, you can target to the customer the kind of service that they deserve: people who spend more money get shorter hold times; people who spend less, wait longer. That's what customer relationship management is really all about.

So what's going on here? The call center industry has become enamored of technologies that act like a Band-Aid on structural problems. People ignore issues of training and morale, creating a workforce that's transient, lightly trained and rarely empowered to actually address the needs of the customer. Technology doesn't demand higher pay, better working conditions, a promotion to a more interesting job, more training. Technology doesn't ask impertinent questions, like customers do.

What people also forget, of course, is that the call center industry is made up of thousands of people who are themselves customers, both in the real world, and when it comes to buying tools for their centers. Here's an unpleasant truth to remember: the vendors of call center tools, though nice people all *are trying to sell you stuff.* Constantly and relentlessly. When they tell you what you need, maybe it's true, and *maybe it's not.*

Lately, they've sold you on email interactions and they're working hard to sell you on live text chat, and on IP routing. They've failed to sell you much in the way of workforce management software, which is unfortunate, because that's one of the few things you really do need. And they've sold a lot of you on monitoring and they're pushing that further into training systems based on those monitoring tools.

Remember that they exist only to maintain the constant flow of product into the call center. When you stop buying, they either go away, or think of new things to sell you. Generally, they do not, however, have the best interests of you, or your customers, or your reps, in their minds. That's not their job. That's the job of the call center manager and his or her colleagues.

But the worst thing you can do — and this is now beginning to play itself out on a global scale, with end-user customers in every industry — is pretend that your customers love you because of the technology you put into their hands. Customers don't have loyalty — that's a complete, total myth. You can make it difficult, or expensive, or inconvenient to leave you. But they don't love you, any more than you love the guys who sell you your ACD or long distance service. When you stop pretending, and see things with clear eyes, you will be able to spend less money on trappings, and put resources where it can do the most good: into the people who connect with customers.

Index

Newton's Telecom Dictionary, 19th Edition

This is the best-selling bible of communications technology, continually updated by Harry Newton with thousands of new terms and revised definitions. It's an essential resource for anyone involved in telecom, networks, computing, and information technology.

"Who wants to go, 'Huh?' when the boss is slinging FLAs (four-letter acronyms)?"
–Smartletter

ISBN 1-57820-307-4, **$34.95**

Going Wi-Fi

Make informed decisions about planning and installing 802.11 "wi-fi" networks. This book helps you tackle the wi-fi challenge, whether installing wi-fi within an existing corporate network or setting up a wireless network from scratch in any business. Author Janice Reynolds guides you through everything you need to know to do it right and to avoid pitfalls in security, reliability, and changing standards.

ISBN 1-57820-301-5, **$34.95**

Find CMP*Books* in your local bookstore

www.cmpbooks.com • 800-500-6875 • cmp@rushorder.com